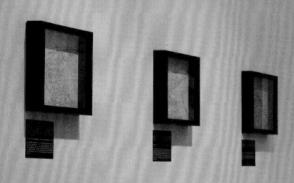

相 遇

2019
上海城市空间艺术季
案例展

《相遇》编委会 编

Edited by *Encounter* Editorial Board

Encounter

2019 SHANGHAI URBAN SPACE
ART SEASON SITE PROJECT

东华大学出版社 · 上海
Donghua University Press Shanghai

主办
上海市规划和自然资源局
上海市文化和旅游局
上海市杨浦区人民政府

HOSTS
Shanghai Urban Planning and Natural
Resources Bureau
Shanghai Municipal Administration of
Culture and Tourism
Yangpu District People's Government

学术委员会

为确保空间艺术季的专业性、学术性和国际性，特成立空间艺术季学术委员会，为空间艺术季各项核心工作提供专业指导和评审意见。委员会体现国际性和专业性，委员均为国际和国内专业造诣高、工作经验丰富且有影响力的专家，并保持国际专家占一定的比例。委员会为多专业的跨界合作，充分体现空间艺术季是专业性、公众性相结合的大型城市公共活动。

职 责
学术委员会职责包括召开各阶段学委会会议，审议有关工作事项，进行策展人和策展方案评审等；活动期间参与论坛和研讨会，进行主旨演讲和学术交流等；出席空间艺术季期间各重要节点活动等。

主 任
本届学术委员会主任由中国科学院院士郑时龄担任。

委 员
学术委员会委员由城市规划、建筑、景观、公共艺术、传播、评论、策展、社会学、媒体等领域专家组成，人数约 30 人。学术委员会委员可根据每届的具体情况做局部调整。

ACADEMIC COMMITTEE

To guarantee that SUSAS is a professional, academic and international event, the academic committee is established to offer professional guides and evaluative opinions on various core tasks of the art season. Committee members, domestic and abroad, feature expertise, experiences and influence, and a considerable of them are internationally renowned figures. The committee is comprised of members from multiple disciplines, fully demonstrating SUSAS as a major urban public event that is both professional and open.

RESPONSIBILITY
The responsibilities of academic committee include: organizing academic committee sessions at each stage where progress, curators and plans are discussed and evaluated; attending forums and seminars during the art season, giving keynote speeches and participating in academic dialogues; attending various key activities during the art season, etc.

PRESIDENT
Zheng Shiling, fellow of the Chinese Academy of Sciences.

MEMBERS
The members of academic committee cover about 30 professionals in urban planning, architecture, landscape, public art, communication studies, art criticism, curation, sociology and press. The list is subject to moderate adjustment according to the requirements of each year.

目 录

Contents

关于 2019 上海城市空间艺术季案例展

2019 Shanghai Urban Space Art Season Site Project

上海城市空间艺术季，继 2015 年、2017 年成功举办两届以来，2019 空间艺术季由上海市规划和自然资源局、上海市文化和旅游局、杨浦区人民政府共同举办，继续以上海的城市更新发展阶段为背景，探讨城市滨水公共空间话题。

上海城市空间艺术季实践案例展是空间艺术季重要组成部分之一，是将实践成果引入展览，将展览成果运用到实践的重要路径，是以点带面，体现各区的积极响应，展示各区近几年来的优秀实践成果，凸显案例"在地性"特色之所在。

2019 实践案例展于 2018 年 5 月启动申报工作，以"滨水空间为人类带来美好生活"为主题方向，结合上海各类滨水空间近年来的城市更新成就，以在地办展的形式举办活动，经过多次踏勘与会议推进，共收到 13 个区的案例申报，经空间艺术季学术委员会会议审议，最终确定 7 类 13 个实践案例展参展名单。

本届实践案例展涵盖了上海各类滨水空间：黄浦江沿线有浦东新区的东岸望江驿、徐汇区的西岸、闵行区的紫竹滨江；苏州河沿线有普陀区滨河园区、长宁区滨河绿带、虹口区北外滩；骨干河道有静安区的彭越浦河岸景观改造；新城水系有青浦区环城水系公园、松江区华庭湖、奉贤区上海之鱼；水乡村庄有金山区漕泾镇水库村……比较完整地展现上海江南水乡各种滨水空间类型，在艺术季期间为公众提供丰富的体验。

In 2019, Shanghai Urban Space Art Season (SUSAS), after its successes in 2015 and 2017, is co-hosted by Shanghai Urban Planning and Natural Resources Bureau, Shanghai Culture and Tourism Bureau and People's Government of Yangpu District. SUSAS 2019, whose background is still the urban generation and development of Shanghai, discusses the topics about waterfront public spaces in cities.

Site projects is an essential element of SUSAS, as well as an important approach to introduce practical accomplishments into exhibitions and apply the results of exhibitions to the real world. These focal points are able to promote the development of broader regions, and as an indication of the responsiveness of districts in Shanghai and a demonstration of the recent excellent practical accomplishments in each district, they highlight the locality of the exhibited cases.

The site project application procedure for SUSAS 2019 whose theme is "How Waterfronts Bring Wonderful Life to People" began in May 2018. The site projects should be related to recent urban waterfront regeneration in Shanghai, and would be held at the actual sites. After multiple on-the-spot surveys and meetings, applications from 13 districts were received. According to the evaluation of Academic Committee of SUSAS, 13 site projects in 7 categories were eventually selected.

The site projects of SUSAS 2019 cover all kinds of waterfronts in Shanghai: along Huangpu River are East Bund river view service stations in Pudong New Area, West Bund in Xuhui District and Zizhu National High-tech Industrial Development Park in Minhang District; along Suzhou Creek are M50 Creative Park in Putuo District, Waterfront Green Belt in Changning District and North Bund in Hongkou District; along another main stream is Pengyuepu Waterfront Landscape Renovation in Jing'an District; at water systems of new cities are Qingpu Round-city Water System Park, Huating Lake in Songjiang District and Shanghai Fish Park in Fengxian District; and there is the water village of Shuiku in Caojing Town, Jinshan District. The above mentioned and other site projects depict a comprehensive picture of the various waterfront spaces in Shanghai, a Jiangnan city of water, and offer rich experiences to the public during the period of SUSAS 2019.

序1
空间赋能 艺术建城——以空间艺术季推动人民城市建设的上海城市更新实践

Preface 1
Energized by Space, Built by Art—Shanghai Urban Renewal Practice Promoting the Construction of People's City with SUSAS

徐毅松
上海市规划和自然资源局局长

Xu Yisong
Director of Shanghai Urban Planning and
Natural Resources Bureau

2019 年 11 月 2 日，习近平总书记深入上海杨浦滨江等地，就贯彻落实党的十九届四中全会精神、城市公共空间规划建设和社会治理等方面进行调研，特别在视察杨浦滨江时做出"文化是城市的灵魂""人民城市人民建，人民城市为人民"等重要指示。以杨浦滨江地区空间改造为代表的上海城市更新实践，在规划理念上注重空间赋能，充分挖掘和发挥空间自然要素附加的生态价值、人文价值和经济价值；在工作实践中注重以点带面，以空间艺术季这类"文化事件"有效激活空间更新的持续活力；在实施机制上注重众筹共治，通过统筹政府、市民和市场等多主体共同协商，不断提升市民在人民城市建设过程中的参与感，不断增强市民对城市空间改善的获得感和幸福感，不断提高规划资源在现代化国际大都市空间治理领域的能力水平。

以"空间艺术季"为载体向滨江空间赋能

新时代的规划资源管理，始终以国土空间源头治理推动城市高质量发展和引导人民高品质生活为己任。正是在这样的理念引导下，上海在城市更新工作中尤其注重空间之于城市生态、地区经济及社会人文的载体作用，并开展了一系列内涵丰富的城市更新实践活动。上海城市空间艺术季（英文简称"SUSAS"）就是其中之一。

艺术季继承发扬"城市，让生活更美好"的世博精神，虽然以展览为形式，但通过空间

On November 2, 2019, General Secretary Xi Jinping went to Yangpu Riverfront in Shanghai to research the implementation of the spirit of the Fourth Plenary Session of the 19th CPC Central Committee, urban public space planning and construction, social governance, etc.. During his inspection of Yangpu Riverfront, he made a few important instructions, such as "Culture is the soul of the city", "People's city should be built by the people and for the people". Represented by the space reconstruction of Yangpu waterfront area, Shanghai's urban renewal practice focuses on energizing the city with space in its planning concept, giving full play to the ecological, humanistic and economic values of the natural space; in the work practice, it effectively activates the continuous vitality of space renewal with such "cultural events" as SUSAS; as for its implementation mechanism, by coordinating the joint power of multiple subjects such as the government, citizens and the market, it continuously enhances the participation of citizens in the process of building the people's city, increases people's sense of acquisition and happiness for the improvement of urban space, and improves resource planning in the field of space governance of a modern international metropolis.

Energize the Riverfront Space via SUSAS

In the new era of planning and resource management, it is always our mission to promote high-quality urban development and provide people with high quality of life with the source management of national land and space. It is under such a philosophy that Shanghai pays special attention to the importance of space as a carrier of urban ecology, regional economy, and social humanity in the process of urban renewal, and has carried out a series of rich urban renewal practices, one of which is the Shanghai Urban Space Art Season (SUSAS).

SUSAS inherits and carries forward the Expo 2010 spirit of "Better City, Better Life". Although it takes the form of an exhibition, it gives more connotation and value to urban space

艺术布展与城市有机更新实践的相互推动，赋予并展示了城市空间更多的内涵和价值。空间艺术季每两年举办一次，活动内容由空间场馆改造主题、实践案例展、地区联合展、公众活动（SUSAS学院）等版块组成，旨在实现"举办一届展览活动、传播一次文化热点、留下一批大师作品、美化一片城市空间"的目的，打造"永不落幕的世博会"。目前连续举办的三届展览活动均选址于黄浦江沿岸地区，伴随"人民之江"建设和滨江地区更新进程，对滨江空间结构优化和品质提升发挥了积极作用。

浦江两岸与上海城市发展历程相戚相伴，拥有丰富的历史风貌遗存，是上海经济、社会、人文内涵丰富的展示窗口。20世纪末以来，随着城市产业结构调整、传统制造业转移以及内港外迁，浦江两岸的工业、码头功能已不适应时代发展需求，也出现了一系列消极空间的问题。因此，滨江地区的城市更新，一向是上海城市更新工作的重点和难点。因此，上海一方面积极推进滨江地区更新规划的编制实施，另一方面通过两年一度的"空间艺术季"，系统地推进和展示滨江城市更新和公共艺术实践，不断对滨江公共空间资源整治赋能。

从2015年室内展示规划成果和公共艺术品，到2017年实现工业遗产文化重塑，再到2019年走出场馆，以城市公共空间为载体，通过与城市更新和公共活动的深度融合，滨江"工业锈带"逐渐转变为"生活秀带"，浦江两岸作为上海全球城市功能品质标杆的形象逐渐清晰，空间艺术季活动也在内涵、深度、广度上不断拓展。三届主展分别选址于徐汇西岸飞机库、浦东民生码头8万吨筒仓、杨浦滨江毛麻仓库及滨江5.5公里的范围内。

徐汇西岸飞机库如今已成为民营美术馆，与西岸其他工业遗产改造的艺术场馆集聚形成了一条艺术品产业链走廊；浦东8万吨筒仓曾是亚洲最大的粮仓，极具历史和文化价值，结合贯通工程进行建筑改造，注入文化休闲、创意展示等多元功能，重新定义了滨江公共建筑和公共空间，在艺术季结束后已举办多

through the cooperation and interaction between the exhibition and urban renewal practices. The Space Art Season is held once every two years and consists of the main exhibition of space venue renovation, site projects, regional joint exhibitions, and public activities (SUSAS College). Each SUSAS is held to create a hot cultural topic, preserve the artworks made by masters at home and abroad, beautify the urban public space and make a "World Expo that never closes". Up to now, three exhibitions have been successfully held along the Huangpu River. Along with the construction of the "People's River" and the renewal process of the riverside area, these exhibitions have played an active role in the optimization and quality improvement of the spatial structure of the waterfront.

Both sides of the Pujiang River are closely related to Shanghai's urban development. With plenty of historical heritages, these places are the showcases for Shanghai's economy, society, and culture. Since the end of the last century, with the adjustment of urban industrial structure, the transfer of traditional manufacturing industries, and the relocation of the inner port, the functions of industries and docks on both sides of the Pujiang River have become unsuitable for the development needs of the times, and a series of space problems have emerged. Therefore, the renewal of the riverfront area has always been the focus and difficulty of Shanghai's urban renewal. Thus, Shanghai has been actively promoting the implementation of the riverside area renewal plan while displaying the achievements of renewal and public art practices through SUSAS, continuously contributing to the renovation of riverside public space resources.

From the indoor display of planning results and public artworks in 2015, to the realization of industrial heritage cultural reshaping in 2017, and then to the deep integration of urban public space with urban renewal and public activities in 2019, the riverfront "rust belt" has gradually transformed into a "show belt", both sides of the Pujiang River have gradually become the benchmark of Shanghai's global urban functional quality, and the activities of SUSAS have been more significant, sophisticated and inclusive. The main exhibitions of the three sessions were respectively located in the previous Longhua Airport airplane hangar of Shanghai West Bund Museum, Xuhui District, the 80,000-ton silo at Pudong Minsheng Port, the Maoma (linen and wool) warehouse with the 5.5-kilometer Yangpu waterfront public space.

The previous Longhua Airport airplane hangar of Shanghai West Bund Museum has now become a private art museum, forming an artwork industry chain corridor with other art venues converted from industrial relics of West Bund. The 80,000-ton silo at Pudong Minsheng Port was once the largest grain silo in Asia. It is of great historical and cultural value. After being restructured, the silo now can be a place for citizens to spend their leisure time and for creative display, making the riverfront

场时尚、先锋艺术活动，成为深受欢迎的艺术空间；2019 年的城市空间艺术季用上海船厂的两座巨大的船坞和具有百年历史的毛麻仓库作展场，再次带领观众领略空间和时间的魅力。杨浦滨江 5.5 公里的岸线上，每件公共艺术品都试图与场地融为一体，成为地景艺术，从静态装置展示到有动态参与的公共空间，实现了历史空间与现代人文的"相遇""对话"。可以说，空间艺术季活动通过对公共空间的赋能，将建筑空间改造、地区更新、视觉艺术设计和社会活动公众参与完整融合在一起，其本身就是现代城市更新理念的一次生动演绎。

以"艺术季"为触媒激发城市更新多重实践

"空间艺术季"作为城市经营的"文化事件"，不仅作用于空间艺术季展区所在地，其"触媒"和"标杆"作用更是激发了滨江 45 公里岸线的整体贯通和品质提升，其"生态、经济、人文"一体的空间赋能理念，也进一步带动了上海城市更新工作的全面展开和品质对标。

"滨江 45 公里贯通"打造世界级滨水功能带

根据黄浦江沿岸地区建设规划，两岸共十个主题区段，打造工业文明、海派经典、创意博览、文化体验、生态休闲、艺术生活等不同主题特色。杨浦滨江段利用老工业遗存更新改造，以工业传承为核心，打造历史感、生态性、生活化、智慧型的滨江公共空间岸线，滨江腹地新建大型办公、商业设施，完善地区功能，为周边市民服务。北外滩地区，位于虹口区南部，与陆家嘴、外滩隔江相望。作为《上海市城市总体规划（2017—2035年）》里国际航运中心功能的核心承载区，在建筑总量基本维持不变的前提下，着眼更大区域联动，致力打造成为卓越全球城市中央活动区的标杆之一。徐汇滨江段，从煤码头和散货码头的集聚区，通过一系列创意产业项目和特色文化场馆运营，已成为全市文化创意产业集聚地；浦东滨江段规划建设世博文化公园，利用 2010 上海世博会场馆区

public architecture and space more than its former usage. It has held several fashionable and pioneering art events after the art season, becoming a popular art space. The 2019 SUSAS uses the original Shanghai Shipyard's two huge docks and its 100-year-old Maoma warehouse as exhibition venues, once again leading the audience into the charm of space and time. In the 5.5-kilometer Yangpu waterfront public space, each public artwork tries to integrate with the site and become a landscape. From static display to a public space with dynamic participation, SUSAS realizes the "encounter" and "dialogue" between historical space and modern humanities. Through energizing the riverfront public space, SUSAS fully integrates architectural space renovation, regional renewal, visual art design, and public participation in social activities, being a vivid interpretation of the modern urban renewal concept.

"Art Season" as a catalyst to stimulate multiple urban renewal practices

As a "cultural event" of urban governance, SUSAS not only improves the area where the exhibitions locate but also serves as a "catalyst" and "benchmark" for the overall coherence and quality improvement of the 45km waterfront area. SUSAS promotes balanced ecological, economic, and humanistic spatial energizing progress and has further driven the comprehensive development of Shanghai's urban renewal and quality benchmarking.

"45 kilometers Joined Up Along the Riverside" to Create a World-class Waterfront Area

According to the construction plan for the Huangpu Riverfront area, there are a total of ten themed sections on both sides of the strait, creating different thematic features including industrial civilization, Shanghai classics, creative expo, cultural experience, ecological leisure, artistic life, etc.. The Yangpu Waterfront section makes use of the old industrial remains to renew and transform, with industrial inheritance as the core, to create a historical, ecological, life-oriented, and smart riverside public space. New large-scale office and commercial facilities are built in the riverside hinterland to improve the regional functions and serve the surrounding area. The North Bund area is located in the south of Hongkou District, facing Lujiazui and the Bund across the river. As the core bearing area of the shipping centre based on *Shanghai Master Plan (2017 – 2035)* international, on the premise that the total building volume remains basically the same, it is committed to building the North Bund into one of the benchmarks of outstanding central activity areas of global cities. The Xuhui Riverside section, through a series of creative industry projects and the operation of cultural venues, has transformed from a gathering area of coal wharf and bulk cargo wharf to the city's cultural and creative industry gathering place. It is planned to build the World Expo Cultural Park on the basis of the 2010

域，总用地面积约 188 公顷。地块位于黄浦江转折的临江界面，堪称"小陆家嘴"，但为延续世博精神、提升中心城区空间生态品质，放弃了近千万平方米开发量，保留下宝贵的绿地，同时将文化与生态结合，配套建设上海大歌剧院等高等级文体设施。这些区域虽然分属不同行政区，但共同遵循"空间艺术季"传播的空间赋能理念，秉承"人民之江"建设总体目标，最终连点成线，共同打造全球城市"会客厅"和世界级滨水文化功能带。

"四大行动计划"引领城市公共空间更新活动

按照"卓越全球城市"的总体发展要求，除积极推进浦江贯通等重大公共空间建设外，上海也考虑市民的多样化活动需求，关注零星地块、闲置地块和小微空间的品质提升和功能创造，将"空间艺术季"理念引入社区，开展持续的城市更新"四大行动计划"，不断增加公共空间的面积和开放度，提高公共空间覆盖率和品质。其中社区共享计划关注老百姓的生活品质问题，重点改造社区消极空间；创新园区计划关注产业地区转型趋势，重点促进产城融合；魅力风貌计划关注历史文脉的保护传承难点，重点探讨历史遗存的活化利用；休闲网络计划则关注市民的健康和休闲需求，提供各种高品质公共空间场所。近年来城市更新项目已遍布全市，类型多样，包括复兴历史建筑和街区的长宁上生新所、静安新业坊；延续历史文脉的长宁愚园路、徐汇岳阳路、南汇新场古镇；改善社区消极空间、改造大量老旧小区的静安彭越浦河；产城融合的张江科学城、市北工业园区等。可以说，以空间艺术季为发端，以黄浦江滨江带为轴线，以"四大行动计划"铺展，上海已形成了不间断、全覆盖、高品质公共空间城市更新的良好局面。

以"共建共治共享"激励公共空间开发多重动力

在空间艺术季的策划和实施过程中，我们积极统筹政府、市场、市民三大主体，建立了贯穿全程的市场与公众参与机制，把空间艺

Shanghai World Expo venues along the Pudong Riverside section, with a total land area of about 188 hectares. The turning sector of the Huangpu River, being called "Little Lujiazui", abandons further development and preserves the green land in order to carry on the Expo spirit and improve the ecological quality of the central urban area. Under the tenet of culture integrating with ecology, high-grade cultural and sports facilities such as the Shanghai Grand Opera House have been built in this sector. Although these areas belong to different administrative districts, they jointly follow the SUSAS concept of energizing the city with space, uphold the overall goal of constructing a "People's River", and finally together create a global city "parlor" and world-class waterfront cultural belt.

"Four Action Plans" to Lead Urban Public Space Renewal

Under the overall development requirements of the "Global City of Excellence", in addition to actively promoting the construction of major public spaces such as the Pujiang Riverside, the diversified activity needs of citizens should also be considered. Shanghai pays attention to the quality improvement and functional creation of fragmented plots, unused plots, and small and micro spaces, introduces the concept of "Space Art Season" into communities, and carries out the "Four Action Plans" for continuous urban renewal, to continuously increase the area and openness of public space and improve its coverage and quality. Among the four plans, the Community Sharing Plan focuses on people's life quality and renovates negative community spaces; the Innovation Park Plan is centred on the transformation trend of industrial areas and promotes the integration of industries and cities; the Charming Landscape Plan concentrates on the difficulties of preserving and passing on historical remains and explores the revitalization and utilization of historical relics; the Leisure Network Plan is targeted on the health and leisure needs of the public and provides various high-quality public places. In recent years, urban renewal projects have been spread all over the city with various types, including the revitalization of historical buildings and neighborhoods in Changning Columbia Circle and Jing'an Xin Ye Fang; the continuation of historical heritage in Changning Yuyuan Road, Xuhui Yueyang Road, and Nanhui Xinchang Ancient Town; the improvement of negative community space and transformation of a large number of old neighborhoods in Jing'an Pengyuepu River; the integration of industry and city in Zhangjiang Science City and Industrial Park in the North Shanghai, etc. With the Space Art Season as a starting point, the Huangpu River waterfront as the axis, and the "Four Action Plans" as a strategy, Shanghai has formed a good situation of uninterrupted, full-coverage, high-quality public space urban renewal.

术季的举办，演化成了引导全社会共建、共治、共享城市更新成果的实践过程。

政府层面

市级政府统筹把控整体建设方向。由市领导牵头，市级各相关部门和各区政府组成联席会议机制，出台《关于提升黄浦江、苏州河沿岸地区规划建设工作的指导意见》作为规划建设纲领性文件，共同研究制定相关政策，讨论审议重要地区规划建设方案，协调解决重大问题。由市级规划资源部门深化城市设计专项研究，加强对天际轮廓线、色彩、公共空间、地下空间等方面的全局管控，将相关要求落实到附加图则中；细化建管要求，充分运用三维审批等手段，加强重点区域项目管理，为沿岸建筑空间品质提供保障。由区级政府主导推动实施协调，明确实施项目、实施主体、实施策略和时间要求，保障规划有序实施。

市场层面

充分调动企业积极性，引导多元主体共同参与建设。为引入多元主体共同参与开发，上海推出一系列适应市场的城市更新政策。比如打破仅政府收储进行改造的单一路径，以释放更多的公共设施和公共空间为前提，鼓励物业权利人按规划进行更新改造；比如实施"带方案"招标挂牌复合出让等，引导和激发市场主体不断提高设计和建设品质。在实施过程中始终算好空间账、经济账和时间账，通过科学的开发时序和目标统筹，平衡好滨江第一层面开放和腹地开发之间的互动，协调好城市历史遗存保护和活化利用的价值，处理好当前投入和长远收益的关系，将资金和市场共同引导到政府对滨江地区长远发展的战略理念上来。

社会层面

引入多样化的公众参与空间治理机制。改变传统规划资源管理理念，在公众参与的深度上，不再仅限于规划编制阶段草案的公示，而是贯穿城市规划管理实施的全过程，公众不再局限于被动的意见征询，而是通过定期举办城市空间艺术季这样的特色活动，让市

"Co-construction, Co-governance and Sharing" to Stimulate Multiple Dynamics of Public Space Development

In the planning and implementation of the Space Art Season, we have actively coordinated the three main bodies of the government, the market, and the citizens, established a market and public co-participation mechanism throughout the whole process, and developed the art season into a practical process of guiding the whole society to the co-construction, co-governance and sharing of urban renewal.

The Government

The municipal government coordinates and controls the overall construction direction. Led by municipal leaders, all relevant municipal-level departments and governments at the district level form a joint meeting mechanism, jointly study and formulate relevant policies, discuss and consider plans for important areas, and coordinate and resolve major issues. The *Guidance on Enhancing Planning and Construction Work in Areas Along Huangpu River and Suzhou Creek* has been issued as a programmatic document. The municipal-level planning and resource departments pay efforts to deepen special studies on urban design, strengthen control on skyline contours, colours, public spaces, and underground spaces, and implement relevant requirements into additional plans; refine construction and management requirements, make full use of technical measures such as 3D approval to strengthen project management in key areas and provide guarantees for the quality of spaces along the shoreline. The district-level government takes the lead in promoting implementation and coordination, clarifying the projects, implementation subjects, implementation strategies, and time requirements, and ensuring the plan is carried out orderly.

The Market

Fully activate enterprises and lead all levels of the society to participate in the construction together. Shanghai has introduced a series of urban renewal policies that are adapted to the market. For example, stop relying only on government acquisition and storage for reconstruction, release more public facilities and public space and encourage property right holders to carry out renewal and renovation according to the plan; and the scheme-based bid-quotation assignment approach which aims to guide and stimulate market agents to continuously improve their design and construction quality. In the implementation process, fully take the spatial, economic, and time cost into account. Through scientific development sequence and target coordination, balance the interactions between riverfront open spaces and hinterland developments, coordinate the protection and utilization of urban historical heritage, deal with the relationship between the current investment and long-term benefits, make the capital and the market serve for the government's strategy of the riverfront area long-term development.

民感知并参与到城市的更新改造中；在公众参与广度上，公众参与的形式也是多种多样的，比如激发社区自治开展的微更新，发动全民为上海的城市更新建言献策；比如建立社区规划师制度，引入专业力量扎根社区；比如与社区管理联合，在公共空间中设立党建服务站，鼓励市民参与公共空间志愿者服务等，让广大市民在城市空间的建设、使用、管理过程中，获得与城市共建共治共享的参与感和成就感。

结语

近六年来，空间艺术季活动已取得了一定的影响力，办展主旨及独特的形式，得到了国内外业界的高度关注和十分积极的评价，已初步形成了一定的城市品牌效应。在今后的城市更新的实践过程中，上海将继续举办好空间艺术季，一是进一步打造城市空间新样板，推动"人民城市"建设。以改善和提升空间品质为根本深入开展城市更新，为人民群众不断创造美好的城市公共空间和生活环境，提高公众参与度和覆盖面，增强群众获得感；二是进一步提供城市发展新动能，放大活动乘数效应。统筹好政府、社会、市民三大主体，调动各方参与城市有机更新的积极性，以城市空间品质提升和文化艺术事件经营的叠加形成乘数效应，促进区域经济和文化密度的提高；三是进一步拓展展览活动新主题，扩大活动辐射影响力。创新工作思路，丰富活动形式和内容，在活动主题上，将从滨水空间逐步拓展到其他各类公共空间；在活动区域上，也将从城市中心地区逐步拓展更贴近市民的社区以及郊野乡村地区。不断提升空间艺术季的内涵，扩大活动的深度、广度和影响力，为建设更有魅力、更有活力、更有温度的人民之城，贡献规划资源的智慧和力量。

The Society

Introduce diversified mechanisms for public participation in space governance. Change the traditional concept of planning resource management. Public participation should no longer be limited to the publicity of the draft plan at the planning stage but runs through the entire process of implementation. Apart from being consulted passively, citizens can perceive and participate in the renewal and transformation of the city through special events such as SUSAS. Public participation should be achieved in various ways, such as encouraging the community to make micro-renewal, mobilizing citizens to offer suggestions for Shanghai's urban renewal; establishing a community planner system, and introducing professionals to work in the community; in cooperation with community management, setting up service stations of the Party in public spaces, encouraging citizens to participate in public space volunteer services, etc. As a result, the public can possess a sense of participation and accomplishment of co-construction, co-governance, and sharing in the process of space construction, use, and management.

Conclusion

In the past six years, SUSAS has become quite influential. The main purpose and special format of the exhibition have received high attention and praises from the art industry at home and abroad, and have initially created a certain urban brand effect. In the future, we will continue to make efforts to hold each SUSAS successfully. Firstly, to create a new model of urban space, and promote the construction of "people's city". We will improve and enhance the quality of space as the fundamental to carry out in-depth urban renewal, continue to create a better urban public space and living environment for the people, increase public participation and its coverage, and enhance the people's sense of gain. Secondly, to provide new momentum of urban development and amplify the multiplier effect of activities. Coordinate the government, society, and citizens, mobilize all parties of the society to participate in the city renewal. Form a multiplier effect with the quality improvement of urban space and cultural and artistic events, to promote the improvement of regional economic and cultural density. Thirdly, to develop new themes of exhibition activities and increase the influence. Be more creative at work and enrich the form and content of activities. The focus of our exhibition will gradually transform from the waterfront space to other types of public space, while the area of activities will also be expanded from communities close to the public to the countryside rural areas. We will continue to improve the connotation of the space art season, hold various activities with more depth, breadth, and influence, and make full use of our planning resources to build a more attractive, more vibrant, and warmer city for the people.

序 2
打响上海文化品牌 助推城市品质提升

Preface 2
Build Shanghai's Cultural Brand, Boost Urban Quality Improvement

上海市文化和旅游局
Shanghai Municipal Administration of Culture and Tourism

文化是城市的灵魂，上海是文化的"码头"，更要做文化的"源头"。设立城市空间艺术季，既是在源头上不断激活和推动文化创新，又是展现城市品位、提升城市品质的重要举措。创立 5 年来，空间艺术季将创造和谐的人居环境、营造宜人的艺术氛围、以城市更新助力文化传承与创新、推动城市公共艺术发展作为目标，已逐渐形成了以激活城市空间的艺术综合展演为形式表征，以关注日常的人文精神表达为美学特征，以深度参与的交互互动为亲民体验特点，日趋成为新的上海文化品牌。

本届空间艺术季以杨浦滨江为主展场，既是落实《全力打响"上海文化"品牌加快建成国际文化大都市三年行动计划（2018—2020 年）》中关于"实施'海派城市地标'品质提升计划，建成特色鲜明的滨江文化长廊"的工作要求，也是基于城市自身发展进程的客观需要。见证上海由近现代民族工业发展的城市工业遗存转换为新的文化综合体，需要从内容的建构上注入更为丰盈的时代文化艺术元素。在前两届的基础上，本届空间艺术季在整体内容架构、公共艺术作品设计、展陈形式、关注观众体验等各个方面都呈现出新的亮点。

艺术之美，塑造城市个性魅力

空间艺术季的内容架构是上海文化发展视野的脚注，既有基于杨浦区域特质、注重文脉梳理的作品，又不乏国际视野的活力展现。

Culture is the soul of the city. Shanghai is the "dock" of culture, and more importantly, it should be the "source" of culture. Holding the Urban Space Art Season is not only to fundamentally activate and promote cultural innovation, but also an important measure to display and improve the cultural environment of the city. In the past five years since its establishment, the Space Art Season has targeted to create a harmonious living environment, forge a pleasant artistic atmosphere, advocate cultural inheritance and innovation with urban renewal, and support the development of urban public art. It has gradually become a new Shanghai cultural brand with a form of a comprehensive art exhibition that activates urban space. It has been dedicated to delivering a kind of daily spiritual culture and enabled the public to enjoy deep participation and interaction with art.

This year's SUSAS takes the Yangpu Riverfront as the main exhibition venue. The *Three-year Action Plan (2018-2020) to Speed Up the Building of the Brand of "Shanghai Culture" and an International Cultural Metropolis* makes the requirement to improve the city quality of Shanghai as a landmark. SUSAS 2019 is not only an endeavor to meet the up-mentioned requirement, but also based on the objective needs of the city's development. In order to show the transformation of Shanghai from the urban industrial remains left by modern national industry development to a new cultural complex, this art season needs more cultural and artistic elements of the times from the construction of content. Based on the previous two sessions, this year's SUSAS presents new highlights in the overall content structure, public artwork design, exhibition format, and audience experience.

The Beauty of Art, Shaping the City's Personality

SUSAS is a showcase of Shanghai's cultural development vision, with works both based on Yangpu's regional characteristics and combining with the history and the international perception. The song cycle *Encounter* is displayed at the opening ceremony. It focuses on individual lives and the combing of history, and

开幕展演的合唱套曲《相遇》以个体生命为着眼点，关注历史文脉的梳理，充满人文关怀的温度。作品由七首单曲组成，按时序描绘了杨浦滨江的七个片断，既是一天之中的七个缩影，也代表着杨浦滨江的前世今生；由烟草仓库化生的"绿之丘"多元空间，展示"杨浦七梦"，围绕杨浦的七个关键词"体育、工人、教育、音乐、河流、纺织及消遣"，勾勒出曾经生活、工作在此地的人们，发生过怎样的故事，如今又对此怀有怎样的梦想。而上海街头艺术节开幕系列杨浦专场和世界舞蹈大赛（中国）巡演，则充分体现上海文化发展的国际视野，前者汇集十多个国家的街艺高手与上海艺人在上海国际时尚中心一起交流技艺，营造台上台下欢乐互动的节庆气氛；后者将"舞蹈世界"融入大众的生活，通过舞蹈语言将不同年龄和背景的人们联系在一起。这些动态的艺术展演与静态的公共雕塑作品展相融合，构成动静相宜的节奏，调动观众的多重感官全方位感受与体验艺术，无形之中塑造着城市的个性与魅力。

以人为本，共享城市艺术活力

以人为本，体现在作品以人的情感、人的经历为表现主体，还体现在将观众的参与和互动作为重点。作为文旅融合发展后举办的第一届空间艺术季，通过艺术项目丰富和提升上海旅游的内涵和品位，通过旅游人气扩大空间艺术季的辐射面，相辅相成。主办方设计了"工业遗存体验之旅"线路，观众从本届空间艺术季的主展馆上海国际时尚中心出发，徒步滨江公共空间，探寻工业遗迹变身时尚地标，听建筑和雕塑背后的故事。此次活动还招募百名市民参与艺术家地绘工作坊，在为期1个月的时间里，创作出长达200米的艺术作品《城市的野生》，永久保存在杨浦滨江的开放空间中。同时，在杨浦滨江5.5公里区段内，共有22处文物和历史建筑可以扫码听故事，还开通了"智慧导览感应系统"，12个感应点位陪伴市民观光讲解导览，带来多维度沉浸式的游览体验。

culture, and is infused with humanity. The work consists of seven songs, depicting seven scenes of Yangpu Riverfront in chronological order, which are the snapshots of the day, representing the past and the present of Yangpu Riverfront. The Green Hill, a space developed from the original tobacco warehouse, showcases the "Yangpu Seven Dreams". It is designed around seven keywords of Yangpu: sports, workers, education, music, river, textile, and leisure, depicting the stories of people who used to live and work here and the dreams they have today. The opening series of the Shanghai Street Performance Festival and the World of Dance (WOD) China Tour fully reflect the international vision of Shanghai's cultural development. The Street Art Festival brings together street art masters from more than 10 countries and Shanghai artists to exchange their skills in Shanghai International Fashion Centre, while WOD brings "the dancing world" into the public life, connecting people of all ages and backgrounds through dancing. These dynamic art performances together with the static public sculpture exhibitions mobilize the audience's multiple senses to feel and experience art in an all-around way, invisibly shaping the personality and charm of the city.

People-oriented, Sharing the City's Artistic Vitality

The people-oriented approach is reflected in the fact that the works take the emotions and experiences of mankind as the main expressions and the participation and interaction of the audience as its priority. This year's space art season is the first one held after the integration of culture and tourism development. The connotation and taste of Shanghai's tourism are enriched by the art projects while the influence of space art season is expanded through tourism popularity. The audience can take the "Industrial Heritage Experience Tour", walk along the riverfront public space starting from the Shanghai International Fashion Centre, which is the main exhibition hall of this year's SUSAS, explore the fashion landmarks transformed from industrial relics and listen to the stories behind the buildings and sculptures. The event also recruits 100 citizens to participate in an artist's ground painting workshop to create a 200-meter-long artwork *Wildness Growing Up in the City* over one month. This artwork will be permanently preserved in the open space of the Yangpu Riverfront. In the 5.5km Yangpu Riverfront section, the audience can scan the code to listen to the story of 22 cultural relics and historical buildings. A smart guide system, with 12 induction points for the public sightseeing tour guide, has also been introduced to create a multi-dimensional immersive tour experience.

The Shanghai Urban Space Art Season will further focus on the concept of "People's City", closely follow the times, and keep up with urban development. Through unremitting creativity and innovation, we will continuously enrich and enhance the brand

上海城市空间艺术季活动将进一步围绕"人民城市"理念，紧扣时代发展的脉搏、紧跟城市发展的脚步，通过持续不懈的创意和创新，不断丰富和提升品牌内涵，为人民创造更多、更美的艺术空间，组织更丰富、参与性更高的文化活动，满足人民群众对美好生活的向往，打造独具魅力和特色的文化品牌。

connotation, create more and better art space for the people, organize various cultural activities to satisfy the aspirations of the people to live a better life and create a cultural brand with unique charm and characteristics.

滨河建设成果展示

Riverfront Site Projects

金山区漕泾镇水库村实践案例展

Site Project in Shuiku Village, Caojing Town, Jinshan District

王夏娴
上海城市公共空间设计促进中心
2019 上海城市空间艺术季实践案例展项目负责人

Wang Xiaxian
Shanghai Design and Promotion Centre for Urban Public Space, Manager of SUSAS 2019 Site Projects

主办单位：
上海市金山区人民政府
承办单位：
漕泾镇人民政府
协办单位：
世录文化创意（上海）有限公司
策展人：
苏冰、董楠楠
展览时间：
2019 年 9 月 22 日—2019 年 12 月 31 日
展览地点：
金山区漕泾镇水库村

Host:
People's Government of Jinshan District, Shanghai
Organizer:
People's Government of Caojing Town
Co-organizer:
Shilu Culture and Creative (Shanghai) Co., Ltd.
Curators:
Su Bing, Dong Nannan
Time:
September 22, 2019 – December 31, 2019
Venue:
Shuiku Village, Jinshan District, Shanghai

为回应 2019 上海城市空间艺术季"滨水空间为人类带来美好生活"的主题方向，金山区有别于其他案例展以城市型江、河滨水空间作为主要展示空间，选择了河网密布、水系发达、水质优良，自然资源禀赋优越的水库村参展，是空间艺术季从城市走向广袤乡村地区的首次见证。

水库村是上海市首批 9 个乡村振兴示范村之一，近年来着力通过打通断头浜，加强河岸护坡生态化景观化处理，提升田园水乡风光。在为期三个月的空间艺术季系列活动中，水库村集中展示以水为核心的乡村振兴建设成果，并结合"衣、食、住、行、劳"的活动主题，策划开展艺术展览、沙龙论坛和多项主题活动。

水库村简介

水库村位于上海市金山区漕泾镇北部，村域面积 3.66 平方公里，耕地面积 3559 亩，

In response to the theme of SUSAS 2019, "How Waterfronts Bring Wonderful Life to People", Jinshan District chooses Shuiku Village, with its extensive network of rivers and creeks, high-quality water and stock of natural gifts, as a first witness to how SUSAS is extending from urban spaces to the massive rural sphere, and offers a distinctive site project from those of downtown riverfronts.

As one of the first nine model villages for rural revitalization in Shanghai, Shuiku Villages has been devoted to connecting dead-end creeks, conserving and beautifying riverbanks and creating enjoyable sights of aquatic Arcadia. The 3-month SUSAS site project focused on how the village is revitalized through water-centred efforts, and delivers art exhibitions, forums and other themed activities covering various daily-life issues such as clothes, foods, residence, transportation and jobs.

Shuiku Village

Shuiku Village is located in the north of Caojing Town, Jinshan District, Shanghai. The 3.66km² of village, with arable lands of 3,559 mu (237ha), is populated by 1,763 registered residents. It features a dense cobweb of wide and undulating rivers, and hence its name, Shuiku (literally "stock of rivers"). Thirty-three

水库村整体效果图（图片来源：漕泾镇）
Rendering of Shuiku Village plan (source: Caojing Town)

户籍人口 1763 人。水库村因水网密布、纵横交错、河宽漾大而得名。水库村内水质优良，全村共有河道 33 条，总长约 23 公里，水面率接近 40%，河面最宽处达 110 米，村内 70 多个独岛、半岛，呈现"河中有岛、岛中有湖"的特色景象。依托特有的自然资源禀赋，水库村大力发展水产养殖和水稻种植，另有西瓜、柑橘等经济作物。

自 2018 年 6 月起，水库村乡村振兴项目正式启动，村内结合乡村规划师制度试点开展了道路、水系、绿化、农业和人居环境等全方位的规划与提升。水库村在关注生态环境优化的同时，注重公共服务设施的完善和基层治理体系的构建。在严格贯彻"多规合一"工作思路中，漕泾镇有序推进郊野公园概念规划、郊野单元（村庄）规划编制工作，并结合漕泾镇总体规划（2017—2035 年），将水库村规划为以品牌农产品种植为主的北部"溪渠田园"、以生活休闲服务为主的中部"滩漾百岛"以及以原生文化滋养为主的南部"荷塘聚落"三个主题片区。在漕泾郊野公园"一园两片"空间布局带动下，激发其核心区与综合服务节点的内生动力，形成水漾农园亮点展示。

rivers of premium water, about 23km in total length and as wide as 110m, cover nearly 40% of the entire village and, with over 70 isles or peninsulas within the village, demonstrate a unique sight of isles-on-rivers and ponds-in-isles. Based on its particular natural endowments, Shuiku Village enjoys bolstering industries of aquaculture and rice farming. Cash crops such as watermelon and orange are also planted here.

Since its launch of rural revitalization in June 2018, Shuiku Village, supported by the program of rural planners, has piloted comprehensive projects to plan and improve roads, water systems, greening, agriculture and living environments. While paying attention to ecological enhancement, the village also focuses on public facilities and the grassroots governance system. In rigidly following the approach of "multiple planning integration", Caojing Town has been steadily developing the conceptual plans of country parks and country units (villages). In this process and in connection with Caojing Town Master Plan (2017 – 2035), Shuiku Village is designated with three thematic zones: the northern Creek Country focusing agricultural product brands, the central Maze of Islands featuring lifestyle, leisure and service, and the southern Lotus Settlement, a conservation for original cultures. The plan of Shuiku Village, promoted by the Caojing Country Park's program of "One Park and Two Divisions", is expected to activate the inherent dynamics of its core area and integrated service nodes and thus create a highlighted demonstration of aquatic country.

空间艺术季助力水库村乡村振兴

水库村是本届唯一的乡村型实践案例展，注重与水库村自身产业的结合，通过"衣、食、住、行、劳"一系列的活动结合乡村振兴的基础设施和公共服务设施举办，由大型户外公共艺术、展演和工作坊三大版块内容组成，且特别邀请了国内20多位艺术家和设计师驻村创作，通过涂鸦墙绘、雕塑、装置、新媒体、音乐、即兴表演等多元丰富的方式结合主题与空间实景再造乡村文化新景观，努力打造人文乡村品牌，使乡村的在地性和艺术季产生真正的联动。

"水酷"实践案例展与农业的结合

1. 大型户外公共艺术、展演与稻田的结合。

金秋九月，正值水库村内金色稻田美如画，在空间艺术季田园实验区邀请艺术家和设计师因地制宜进行了驻地创作。在稻田中央搭建起了舞台，舞台旁边放置了雕塑家陈剑生的《方舟》等雕塑作品。舞台中央，伴随钢琴现场演奏，巨幅油画由著名画家林加冰现场创作完成，角市的"大龙组合"带来水晶球、手碟和武幻双环组合而成的杂耍串烧，不时赢得观众阵阵喝彩。除此之外，在水泾路长堰路路口的稻田处，还举办了儿童书法教学、现场绘画与表演等。

2. 艺术季的田园野趣活动。

案例展活动选取水库村沈家宅南侧的生态池塘为野趣体验区，结合村内市级重点现代农

田园油画创作（图片来源：漕泾镇信息中心）
Painting show among paddies (source: Caojing Town Information Centre)

现场书法作品（图片由漕泾镇信息中心提供）
Calligraphy show (source: Caojing Town Information Centre)

How SUSAS Promotes Rural Revitalization in Shuiku Village

Shuiku Village offers the one and only rural site project of SUSAS 2019. The programme integrated with local trades consists of three panels – large outdoor public art, performance and workshop – in connection to the infrastructure and public facilities as part of rural revitalization, and activities covering issues such as clothes, foods, residence, transportation and jobs. More than 20 domestic artists and designers are invited to settle in the village and create together a new landscape of thematic and authentic rural cultures in a variety of forms such as graffiti, sculpture, installment, new media, music and live performance. It is intended to establish a brand of cultural village featuring real interactions between rural locality and SUSAS.

"Water Cool" Site Project: Agricultural Elements

1. The combination of large-scale outdoor public art, exhibition and performance, rice fields.

儿童赤脚摸鱼活动（图片来源：漕泾镇信息中心）
Young Bare-foot Hand Fishers (source: Caojing Town Information Centre)

Amidst the golden rice paddies in September, artists and designers are invited to contribute their talents in the SUSAS Country Art Experiment Zone in Shuiku Village and create on-site art pieces aligned with local features. A stage is erected in the centre of paddies, accompanied by sculptures such as *Arch* of Chen Jiansheng. Lin Jiabing, a famous painter, draws a massive oil-painting in the centre of the stage along with the tunes played by pianists. The Dragon Troupe of Jiaoshi arouses bravos with

业项目 (以大型生态池塘为基础进行设施化生态高效流水养殖)，组织了生动有趣的乡野童年赤脚摸鱼、野外生存小渔夫活动，邀请小朋友们体验水库村田园艺术季的野趣，并科普了生态养殖知识。生态池塘通过在池塘沿岸陆基运用设施化工程技术手段、在水中种植景观植物等立体种养结合模式，对养殖产生的有机排放污物进行无害化综合利用，实现内循环零排放，在上海地区属于首创。

"水酷"实践案例展与基础设施和公共服务设施的结合

1. 主题演讲与活动沙龙。

"水 COOL · 2019 乡村艺术季之乡村振兴系列论坛暨第二期·乡村建筑主题论坛"在水库村综合为老服务中心举办。副镇长高楠出席论坛并致辞，同济大学建筑与城市规划学院景观学系副教授董楠楠主持论坛。为老中心占地面积约 7 亩，由长堰路北侧废弃已久的老村委会改造而成，是集"日间照料""长

its acrobatic portfolio of crystal ball juggling, handpan and buugeng. What's more, calligraphy classes for kids, painting show and performances can be found by the paddies at the intersection of Shuijing Road and Changyan Road.

2. SUSAS country joy activities.

The ecological pond at the south of Shenjia Estate, Shuiku Village is selected as a Country Joy Experience Zone where interesting activities, in association with the municipality-level key modern agricultural project in the village (efficient and ecological running-water aquafarming facility based on large ecological ponds), are organized for children to experience country joys of Shuiku Villages and learn about ecology-friendly agriculture as bare-foot hand fishers or wild survivors. These ecological ponds constitute a zero-discharge system of internal circulation, the first of its kind in Shanghai, via a vertical integrated planting-aquafarming approach – engineering and technological facilities on the banks, landscape plants in the water and more – which recycles and reuses the wastes and emissions produced by aquafarming.

"Water Cool" Site Project: Infrastructure and Public Facilities

1. Keynote speech and salon.

"Water Cool 2019 Rural Art Season: Rural Revitalization Serial

《水天一色》立面彩绘作品（摄影：妮科尔）
Sky and Water in Oneness, facade decoration (photograph: Nicole)

者护理""助餐服务"为一体的村民驿站，建筑总面积约 2500 平方米。其中，主体建筑包括北侧两层长者之家和南侧单层老年人日间照料中心、康复室、医务室、护理室、活动室、阅览室、浴室等一应俱全，有 32 个床位供老年人全日制居住。建成后，"日间照料"中心的大厅常常座无虚席，广受居民的欢迎。

Forum and the Second Rural Architecture Thematic Forum" at the Comprehensive Elderly Care Centre of Shuiku Village is attended and addressed by Gao Nan, deputy mayor of Caojing Town. Dong Nannan, associate professor of Department of Landscape, College of Architecture and Urban Planning, Tongji University, acts as the host. The Elderly Care Centre, converted from the abandoned office of village party committee to the north of Changyan Road, covers an area of about 7 mu (4,667m²) and houses functions of day care, elder nursing and canteen in its room of 2,500m². The main buildings include the two-storey

《逍遥游》渔村生活图（摄影：妮科尔）
Carefree Excursion, a depiction of life in a fishing village (photograph: Nicole)

2. 村内艺术作品。

艺术季期间，水库村因为各种艺术作品的装点，分外热闹。艺术作品展示漕泾镇深入挖掘村情、村史，留住乡音、乡愁，延续乡土文化气息，从时代回忆望见村庄新生。以下主要介绍其中三个。

a.《水天一色》+ 宅基农宅更新。

结合水库村村域内"三高"沿线及零星宅基的归并需求，由政府统一建造的一期南片区东临村民中心合计安置农户 42 户，通过点状供地实现建设用地节地率达 50%。几户新建的农宅外立面由艺术家施政画出了一个充满科幻色彩的水星世界，该作品名为《水天一色》，描绘当人在充满水的星球生活时漂浮的状态以及与宇宙的关系，营造了村内艺术季活泼的氛围。

b. 波普装置艺术 + "藕遇" 公园。

艺术家李海滔以波普装置手法结合金山农民画元素，创作了一幅逍遥自在的渔村生活图，结合装置取名为《逍遥游》，放在"藕遇"公园临湖"荷花型"亲水木栈道进行展示。"藕遇"公园位于核心区两条主干道交汇处，三

Elder Home and the single-storey Day Care Centre to its south. A full range of services are provided here: convalescence, clinic, nursing, lounge, library and among others, bathing. 32 beds are available for full-day dwelling. The Day Care Centre is so popular that its hall is frequently filled with locals.

2. Art pieces in the village.

During the period of SUSAS 2019, Shuiku is particularly vibrant for all the art pieces decorating the village. The artwork demonstrates the efforts of Caojing Town to thoroughly evacuate the features and histories of Shuiku Village, preserve local tongues and sentiments,aintain rural cultures and envision the new life of Shuiku Village with the hindsight of its memories. Now we will empathetically introduce three significant art pieces.

a. *Sky and Water in Oneness* + Homestead cottage renovation.

To the effect that scattered homesteads or homesteads along high-speed railway/expressway/elevated way need to be incorporated, 42 households of Shuiku Village are resettled in the first-phase residences at the southern part of the village and to the west of Villager Centre. The residences are all constructed by the government. Thanks to its dotted arrangement, they now take up 50% less land. The facades of several such new houses are decorated with *Sky and Water in Oneness* by the artist Shi Zheng. The visualized science fiction depicts an aqua planet in which everyone floats on water and shows the relationship between humans and the universe, fostering a lively atmosphere in the village during SUSAS 2019.

《人—原野》雕塑作品（摄影：妮科尔）
Man in the Field, sculpture (photograph: Nicole)

面环水，占地面积约 50 亩，由原有藕塘改造而成，公园内荷田碧波荡漾，与村内生态浮岛、多样化的湿地景观交相辉映，西侧废旧小屋则改建为休闲书吧，建筑面积约 490 平方米，公共区域设有中心广场、阶梯式休闲区、"书箱"式观景台，后期计划采用公建民营模式进行长效运营。

c.《人—原野》＋湿地公园。

《人—原野》雕塑作品由德国艺术家周浩南创作完成，雕塑装置运用凝固的轮廓，加上光的折射，展现人与水库村内湿地公园水岸田野风光的融合。湿地公园位于水库村中心河东段南侧，由原有虾塘改造成为敞开式河流湿地，总面积约 106.5 亩，属于市级水利专项施工范围之一。通过堆筑多个小岛屿，种植各类植物，修建连接岛屿的栈桥和漫步栈道，形成水上迷宫。

b. Pop art installation + "Lotus Encounter" park.

Combining the techniques of pop art and elements of Jinshan farmer paintings, Li Haitao composes *Carefree Excursion*, an installation depicting the free lifestyle of a fishing village. It is placed on the lakeside lotus-shaped plank road in "Lotus Encounter", a park at the intersection of the two main streets in the village's core area. The park, surrounded by water on three sides and transformed from a lotus pond, covers an area of 50 mu (3.33ha). The rippling lotus fields complement the ecological floating islands and a diversity of wetland sceneries which the village has to offer. The unused hut to the west of the park is repurposed to be a bookshop of 490m² whose public space features a central square, leisure terraces and a "book case" observation deck. The publicly constructed bookshop is planned to recruit a private operator to keep it running.

c. *Man in the Field* + Wetland park.

Man in the Field, a sculpture created by German artist Roland Darjes, uses its frozen outline and refracted light to demonstrate how humans and the country sceneries on the banks of Shuiku Village wetland park are converged. The Wetland Park of 106.5

根据种植植物不同、功能不同，湿地共分为固土的芦苇荡迷宫、鸟类筑巢的杉林迷宫、涉禽觅食及繁殖的寻鸟秘影迷宫以及游客赏花探果迷宫 4 个区域供人游览。

3. 长堰路提档升级 + 开幕活动长桌宴。

水库村开幕式结合村民中心举办。水库村村民中心占地面积约 15 亩，经对现村委会以及水泾路东侧建筑部分拆除、整合、提升，改造成集村委会办公、为民服务、文化活动为一体的村民中心，内设一门式办事大厅、文化礼堂、老年活动室、便民商店、医务室、监控室、消防室、村史馆以及相

mu (7.1ha) is located at the southern side of the eastern section of the river running through the centre of Shuiku Village. What used to shrimp ponds are now open river wetlands and a designated municipal hydraulic engineering project. Stacked and vegetated islets connected with plank roads or bridges constitute a series of mazes on water. Depending on the types of vegetation and functions, the wetland is divided into four areas open to visitors: Reed Maze on consolidated soils, Cedar Maze for nesting, Bird Maze where waders feed themselves and mate, and Floral Maze with appealing flowers and fruits.

3. Changyan Road Upgrade + Inaugural Long Table Feast.

The opening ceremony is held in connection with the Villager Centre. The Villager Centre covers an area of about 15 mu (1ha), and is transformed from the existing office of Village

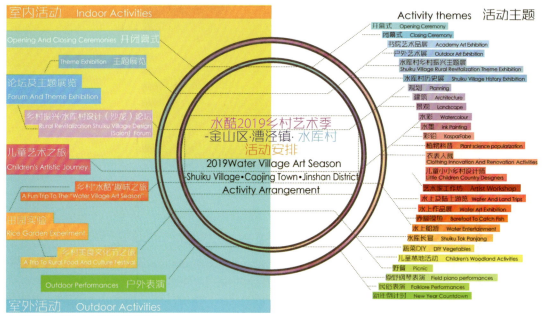

水库村活动一览表（由策展团队提供）
Event programme of Shuiku Village (image: curatorial team)

苏冰（左）与董楠楠（右）
Su Bing (left) and Dong Nannan (right)

金红（漕泾镇规建办副主任）：

在 2019 水 COOL 乡村艺术季期间，让我感受到了乡村与艺术的"相遇"、艺术家与村民的"相遇"，非常感谢设计师和艺术家走进村里，给这个朴素安静的乡村带来了更多的生机活力。在近一年的建设中，各级领导、设计师、工作专班、村委会齐心协力，一起规划设想、讨论方案、走工地、催进度，终于，我们可以漫步长堰路的林间步道、可以泛舟水库中心河、老人们可以在为老中心下棋聊天、第一批村民搬进了令人羡慕的集中居住新居……同时，也非常可喜地看见很多村民热情地参与其中。期待未来的乡村更艺术、更美好、更让人向往！

Jin Hong (deputy director, Planning and Construction Office of Caojing Town):

During the 2019 Water Cool Rural Art Season, I felt the "encounter" of village and art, of villagers and artists. I really appreciate that the designers and artists would come to the modest and quiet village, and energizes it. The development procedure of nearly one year witnesses the combined efforts of governmental leaders, designers, task forces and the village party committee as they envision plans, discuss designs, visit construction sites and boost progress. And eventually, we can wander on the forest lanes of Changyan Road, boat on the central river. Elders can chat and play chess in the Elder Care Centre. The first batch of villagers move into their new admirable concentrated residences…Meanwhile, we are delighted to find that many villagers are enthusiastically participating. Let's hope that the village will be more artistic, more beautiful and more desirable!

苏冰（跨界策展人）：

2019 年金山区水库村田园实验让艺术再次介入乡村。一方水土一方人，我们还是采取了因地制宜的方式，通过在地艺术创作、高校、村镇与广泛的社会资源全面链接合作，以临时性场地和在地化建造技术，导入特色化的艺术活动与文化体验，旨在促进城乡人群与社会资源之间的交流互动，艺术家通过体验、亲近乡村的风俗人文，也更好地激发艺术创作融入日常生活和场景中。

Su Bing (interdisciplinary curator):

The 2019 countryside experiment in Shuiku Village, Jinshan District again injects art into the rural. Since every place and the dwellers it fosters are unique, we have adopted an adaptive approach. While connecting and combining on-site artists, universities, the town, the village and the wider public resources, and based on temporary sites and on-site construction technologies, we have introduced characteristic art activities and cultural experiences in order to facilitate the flow and interactions between city dwellers and villagers, and among various social resources. On the part of artists, approaching and experiencing rural cultures may inspire them to integrate creative art into daily life and scenarios.

董楠楠（水库村乡村规划师）：

在本年度的城市空间艺术季里，不仅呈现了这个都市和水的华丽相遇，而且看到了乡村振兴中新型田园水乡的风采灵韵。衷心感谢苏冰老师的策展协调和城市空间艺术季的资源平台，深深感动于三个月中热情支持我们的各级领导、参与建设和维护的乡亲们、关注和加入本次活动的规划设计专家、艺术家、工程建设单位和各类参与的执行团队。衷心期待在未来的艺术季活动中看到更多更精彩的海派乡村呈现。

Dong Nannan (rural planner of Shuiku Village):

SUSAS 2019 not only exhibits the marvelous encounter of the metropolitan and waters, but also showcases the elegance and energy of new-type water towns in rural revitalization. I really thank Su Bing for his curatorial coordination, and SUSAS for its resourcing platform. I am deeply moved by the leaders at all levels with their immense support, the dear villagers with their participation in construction and maintenance, and all the planners, designers, artists, construction units and everyone else who has attended to and joined our project. I sincerely hope to see more Shanghai villages in future SUSAS.

关旅游服务设施等。

开幕式的重要活动之一，长桌宴结合水库村"四好农村路"提档升级改造，设置在长堰路上，结合厨王争霸赛、河鲜宴等活动，为庆祝新中国成立70周年，摆宴70桌，开启了乡村案例展浓浓烟火气的序幕。作为金山区首条投入施工的"四好农村路"改造项目，长堰路改造工程西起朱漕路，东至水泾路，全长904米，车道宽度7.5米，道路等级提升为三级公路。其主要包括长堰路车行道翻新及局部拓宽、翻挖路缘石、侧平石，与工程同步实施路灯、交通标志标线、绿化等，修缮总面积7405平方米，且通信架空线实施入地改造，将农村生活污水、天然气、自来水管网全部铺设到位。

"水酷"乡村艺术季的展后感言

2019"水酷"乡村艺术季向我们展示了水库村在基础设施与公共服务设施规划、设计、建设的精心与用心，通过大型户外公共艺术、展演和工作坊的多元表现形式，以涂鸦墙绘、雕塑、装置、新媒体、音乐、即兴表演等丰富的演绎方式，结合乡村振兴实践成果，突出水库村季相特点、农时丰收的场景，体现出乡村空间变化的"原味"之美，打造了一幅生动的新江南田园景观。

在筹办2019上海城市空间艺术季水库村案例展期间，漕泾镇领导、项目组成员与策展团队多次召开方案讨论会，在筹备和活动期间，引导当地村民积极参与，最终呈现给大家一种乡村与艺术的"相遇"，艺术家与村民的"相遇"，塑造了乡村形，又留住了乡村魂。让我们共同期待未来的水库村更美更富更强！

Party Committee and the building on the east of Shuijing Road via demolishing, integration and improvement. Incorporating functions of village party committee office, civil services and cultural events, the new Villager Centre consists of a one-door service hall, a lecture hall, elder lounges, convenience shops, a clinic, a monitoring room, a fire station, a village history museum and relevant tourist service amenities.

As an important event of the opening ceremony, Long Table Feast is placed along the upgraded Changyan Road. In addition to Chef Championship and Dinner of Riverfoods, the 70 tables of dishes in the feast, as a tribute to the 70th anniversary of People's Republic of China, make a homely opening for the rural site project of Shuiku Village. As the first completed "Four Well" rural road – well built, well managed, well protected and well maintained – in Jinshan District, the improved section of Changyan Road is 904m long and 7.5m wide, extending from Zhucao Road at the west to Shuijing Road at the east, and is upgraded to be a third class highway. The upgrade project – renovating and partly widening the vehicle lanes, excavating curbs and gutter aprons – and simultaneous installment of street lamps, traffic markings and road planting cover an area of 7,405m². At the same time, the overhead telecommunication lines are buried, and the pipe networks of drainage, natural gas and tap water are also completed.

Testimonials for Water Cool Rural Art Season

2019 "Water Cool" Rural Art Season demonstrates how Shuiku Village has devoted its efforts and considerations to planning, designing and constructing infrastructure and public service facilities through large outdoor public art, performances and workshops. A diversified combination of graffiti, sculpture, installment, new media, music and live performance, in addition to the practical accomplishments of rural revitalization, are utilized to highlight the seasonal and harvesting scenes of Shuiku Village, show the "original" beauty of the transforming rural spaces, and create a vivid scenery of contemporary Jiangnan countryside.

同创水岸新生活——嘉定区实践案例展

Create a New Waterfront Life Together — Site Project in Jiading District

陈成
上海城市公共空间设计促进中心

Chen Cheng
Shanghai Design and Promotion Centre for Urban
Public Space

主办单位：
上海市嘉定区人民政府
承办单位：
嘉定区规划和自然资源局
协办单位：
嘉定团区委、嘉定区建设管理委、嘉定区
绿化市容局、嘉定区文化旅游局、嘉定镇
街道、菊园新区管理委员会、嘉定区国有
资产经营（集团）有限公司、绿洲控股集
团
策展人：
周芳珍
展览时间：
2019 年 10 月—11 月
展览地点：
嘉定环城河步道、现厂创意园区（博乐路
70 号）

Host:
People's Government of Jiading District, Shanghai
Organizer:
Jiading District Planning and Natural Resources Bureau
Co-organizers:
Jiading District Committee of the Communist Youth League
of China, Jiading District Construction and Administration
Committee, Jiading District Bureau of Afforestation and Urban
Environment, Jiading District Bureau of Culture and Tourism,
Sub-district Office of Jiading Town, Management Committee
of Juyuan Sub-district, Jiading District State-owned Assets
Management (Group) Limited Company, Shanghai Oasis
Holding Company
Curator:
Zhou Fangzhen
Time:
October – November, 2019
Venue:
Jiading Moat Promenade, "Now Factory" Creative Office Park
(70 Bole Road)

"同创水岸新生活"是 2019 上海城市空间
艺术季在嘉定古镇以嘉定环城河步道贯通提
升作为项目载体的实践案例展。2018 年环
城河全线贯通并对市民开放，整个环城河滨
水空间从过去封闭的空间变成如今重要的公
共开放空间，展现了水资源的生态环境提升
及城市滨水空间满足人民日常生活的转变。
步道贯通也使环城河影响力得到了提升，现
已成为了嘉定古镇文化标志水岸和新的城市
名片。

嘉定实践案例展通过艺术与文化活动来激发
居民与滨水空间的互动，使居民亲触城市变
化，感受更多的获得感和幸福感，从而呼应
空间艺术季的主旨——"滨水空间为人类带
来美好生活"。

"Create a New Waterfront Life Together" is a SUSAS 2019 site
project at Jiading Old Town for its moat promenade. Since
Jiading Moat Promenade was completed and open to public in
2018, the entire waterfront space around the town, which used
to be closed, has become an important public open space,
showing that waterbodies have improved ecologically and that
urban waterfronts are shifting in their ways to satisfy people's
needs of daily life. The promenade has also improved the
influence of the moat, turning it into a cultural symbol of Jiading
Old Town and a new city name card.

The Jiading site project, with its art and cultural events, activate
interactions between citizens and waterfronts, allow citizens
to personally touch the city's changes and feel more sense of
gaining and happiness – hence the theme of SUSAS 2019, "How
Waterfronts Bring Wonderful Life to People".

嘉定古城护城河与城墙（图片来源："上海嘉定"公众号，2019 年 10 月 18 日）
Moat and wall of Jiading Old Town (source: "Shanghai Jiading" WeChat Subscription Account, 18 September 2019)

环城河步道贯通背景

承袭 800 年的城池布局

嘉定古城建于南宋嘉定十年（1217），距今 800 余年，护城河和城墙是当时古城重要的防御设施。环城河总长 6.5 公里，绕城一圈呈圆形。横沥河、练祁河与环城河纵横出入，南北走向的为横沥河，东西走向的是练祁河，横沥河北接浏河、南通吴淞江，练祁河东入大海。嘉定人形象地称之为"十字河"，它们与环城河形成"十字加环"的独特水系。嘉定建城之初，城内因"十字河"而自然分割成四块片区，数百年来居民多居于河两岸，沿河生产生活，形成了最早的古镇布局。州桥老街、孔庙、法华塔、秋霞圃等众多文物古迹沿河网分布，构成了丰富而独特的历史文化风貌，一个典型的江南水乡跃然眼前。1991 年嘉定镇被评为上海市四大历史文化名镇之一，2008 年被评为中国历史文化名镇。

800 年来，护城河静静地滋养哺育着嘉定代代百姓。她是上海地区唯一保存完好且依旧具备实用功能的古代护城河，至今仍发挥着城市河道和航道的功能，河道沿线历史遗存众多、人文底蕴丰厚、滨河绿带公园相连，是一条具有稀缺性和重要历史意义的河流。

Background of Moat Promenade Connection

Layout Founded 800 Years Ago

Jiading Old Town was established in 1217, more than 800 years ago, when its moat and walls were essential defenses. The circular moat, 6.5km in length, runs full around the town. Hengli River which flows south from Liu River to Wusong River, and Lianqi River which runs east to the sea form, as locals say, look like a "cross", and if we also consider the moat, we have a unique pattern of rivers: Circle and Cross. At its beginning, the town was naturally divided by the "cross" into four quarters. For hundreds of years, most people lived and worked along the rivers, and the original town layout is thus founded. The plethora of historical relics – Zhouqiao Ancient Street, Confucian Temple, Fahua Pagoda, Qiuxia Garden and more – distributed along the rivers constitute a uniquely rich scene of histories and cultures, reminiscent of a typical town in the lower basin of Yangtze River. Jiading Old Town was declared to be one of the four Municipal Famous Historical and Cultural Towns of Shanghai in 1991, and became a National Famous Historical and Cultural Town in 2008.

The moat has silently nurtured generations of Jiading residents for 800 years. As the only intact and functional ancient moat in Shanghai, it is still a navigable urban waterway. The numerous historical relics along the river, its deep cultural significance and the adjacent green belt parks have made it a rare and historically important river.

展现新清明上河图景

2017 年，嘉定区委、区政府将环城河步道规划定位为"申城千年的清明上河图"，滨河改造范围约 12.2 万平方米。设计注重对城市文化历史的承袭，在展现历史赋予城市原有的面貌和气质的同时，从"环通、植绿、河景、乐游"这四方面进行设计，通过统筹两岸的步道、绿化、驳岸、灯光和配套设施等布局，将其打造成为生活休闲水岸、生态公园水岸、历史人文水岸三大风貌段。

总长约 6.5 公里的嘉定环城河道环嘉定老城而建，涉及范围 4.02 平方公里，沿途有企业、高校院所，也有居民小区，涉及的主体众多，人口也相对密集。这也给环城河步道建设制造了不小的"障碍"。

多方合力、攻坚克难实现贯通

环城河步道工程分为内外两圈，从"贯通、安全、美化、便利"四方面着手，新建了北水关桥、练祁河桥、清镜河桥 3 座桥梁，打通了 12 处桥下通道，串联起嘉定紫藤园、南水关公园等多座公园绿地。在推进步道贯

A New Picture of Along the River During the Qingming Festival

In 2017, Jiading District Committee of the Communist Party of China and Jiading District Government designated the moat promenade as "the millennial Shanghai version of Along the River During the Qingming Festival". The renovated waterfront amounts to about 122,000m². The design, while trying to inherit the history and culture of Jiading and demonstrate the features and temperament bestowed upon the town by its history, consists of four aspects: connection, vegetation, waterscape and amusement. The waterfront is intended to be a hybrid of three sections – lifestyle and leisure, ecological park, history and culture – through an integrated plan of pedestrian lanes, greens, revetment, lighting and amenities.

The 6.5km moat promenade runs a full circle round Jiading Old Town and involves an area of 4.02km², within which are a variety of entities such as businesses, colleges, research institutions and apartment complexes, and a dense population. All these are "obstacles" to overcome when the moat promenade is constructed.

Combined Efforts to Overcome Difficulties

The Moat Promenade consists of the Outer Ring and the Inner Ring, and for sake of connection, safety, beauty and convenience, 3 new bridges – North Water Pass Bridge, Lianqi River Bridge and Qingjing River Bridge – and 12 under-bridge

嘉定环城河步道工程
Jiading moat promenade

通过程中，破解工作"难点"，打通沿岸"堵点"，连接步道"断点"，这项工作也得到了沿线企事业单位、院校和社区居民的大力支持。

比如，嘉中社区和李园二村原有的自行车棚临河而建，为保证贯通工程顺利进行，嘉定区相关部门在沿线小区内就车棚的设计方案等展开广泛的民意征询，比较修改后与居民取得了共识。最终，小区的车棚往内部分退让的同时，还成功挖潜了低效的使用空间，新增了部分机动车停车位。又如，上海大学的小区内，也有一段500米的区段处在贯通的关键位置。经过与上海大学校方的反复沟通，此段滨河区域终于实现了全天候开放。为了延续步道的美观，上海大学校方主动提出用绿化来将校区与河道进行软隔断，在确保校内管理相对封闭的同时，也保持步道景观环境的一致性。

环城河步道贯通后，为了方便小区居民的出入，特地在沿线的社区开设了11处便门，并通过门禁系统来实施管理。对于一些确实无法腾挪沿河空间的既有小区，采用增设步行桥和栈道的方法来实现步道贯通。

打造为百姓喜爱的网红空间

2018年7月，步道内圈实现贯通，贯通后深受周边居民群众的欢迎，迅速成为新的"网红"。居民群众真正感受到了嘉定城市建设给日常生活带来的舒适和便利。2019年9月，长约9.5公里的外圈也实现了贯通。贯通后，嘉定将计划推动步道内、外圈联动，构建环城河步道慢行交通系统，拓展市民休闲健身设施。未来，嘉定还将试点水上旅游，推动环城河旅游品牌项目建设。将来市民可以"坐着游船游古城河"，畅想着与历史对话的美好景象。

嘉定在地的实践案例展示

"同创水岸新生活"实践案例展由"河图"当代艺术展、嘉定环城河城市设施/城市家具优秀作品展、嘉定城乡规划展、"嘉定环城河的昨天、今天和明天"图片展4个主题展览，以及高端论坛、环城河步道定向打卡

passageways are built to connect Jiading Wisteria Garden, South Water Pass Park and other public greens. The businesses, colleges and residents along the promenade have actively supported the project to overcome difficulties, break blockages and connect dead-ends.

For example, for sake of the promenade project, the riverside bicycle sheds of Jiazhong Community and Liyuan Second Village had to be dealt with, so the related authorities of Jiading District comprehensively surveyed locals about how to redesign the sheds. Eventually, a consensus was reached: the sheds shall partly withdraw into the communities while additional car parking slots were made available by utilizing underused spaces. The residential quarter of Shanghai University, of whom a section of 500m falls exactly on an essential location of the moat promenade, offers another example. After repeated negotiation with the university, this waterfront section is finally open 7/24. And to prevent discontinuance in scenery, the university proposed to loosely block the campus and the river with plants so that the closed-off campus management is maintained while the promenade may enjoy a consistent appearance.

After the promenade is completed, 11 gates with access control are installed for residents of neighborhoods along the promenade. And for certain existing neighborhoods whose riverside spaces cannot be made available, new footbridges and plank roads are constructed for a continuous promenade.

Create a Popular Internet-famous Destination

The inner ring was completed in July 2018, and it quickly became popular among local residents and famous across the Internet. The project allows the citizens of Jiading to truly feel how urban development has made the town more comfortable and convenient. When the outer ring, about 9.5km in length, was also completed in September 2019, Jiading was planning

《天边那朵云》系列作品 / 艺术家：郑晓辉
Cloud in the Sky by Zheng Xiaohui

《冷水》/ 艺术家：林森
Freezing Water by Lin Sen

《丝路谣》/ 艺术家：李付彪
Song on Silk Road by Li Fubiao

《听》/ 艺术家：朱勇
Listen by Zhu Yong

《峥嵘》/ 艺术家：李宪阳
Lofty by Li Xianyang

"河图"当代艺术展现场
Venue of contemporary art exhibition "River Map"

赛、嘉定环城河步道城市家具设计方案征集评选 3 个系列活动构成。

"河图"当代艺术展

"河图"当代艺术展邀请到了活跃在当代艺术领域的 7 位青年艺术家：李付彪、李宪阳、林森、庞海龙、于洋、朱勇和郑晓辉。艺术家们带来了近期创作的 10 多件佳作，并将这些多元化的艺术作品置入到环城河步道的绿带空间里，营造出更加接近"生活本身场景"的艺术空间效果和体验。

此次艺术展着力于人与自然的和谐，用多种艺术形态的作品给人们带来耳目一新的新体验、新感觉。其在城市空间生态环境中形成一种传统与当代艺术对话的氛围，以别样的视角激活城市空间的文化生态。当市民进入环城河的滨水空间，也像是走进一场丰富多彩的文化之旅，唤醒一种更为深刻的内心感悟。

环城河城市设施／城市家具优秀作品展

环城河城市设施／城市家具优秀作品展以"嘉定环城河城市家具设计方案征集"为基础，并邀请若干新锐设计师，以嘉定文化为主线展开创作，从公共艺术的角度对城市设施、城市家具进行再设计，并以展板及实物的形式呈现。

嘉定城乡规划展

嘉定城乡规划展以嘉定区总体规划（2017—2035 年）为基础，充分展示嘉定未来城乡空间的规划发展愿景，即到 2035 年把嘉定建设成为上海大都市圈的现代化新型城市，打造汽车嘉定、科技嘉定、教化嘉定、健康嘉定、美丽嘉定；紧密结合区委区政府提出的"乡村振兴"战略目标，展示嘉定全面推进各涉农镇郊野单元规划的主要成果。

"环城河的昨天、今天和明天"图片展

嘉定"环城河的昨天、今天和明天"图片展在嘉定环城河两侧公共空间、现厂创意园区（博乐路 70 号）举办，以《嘉定，一座城池的前世今生》（主编，甘永康）一书为基础，通过展示摄影作品图片的形式，让市民

to connect the rings and build a full slow-traffic system around the moat, expanding the leisure and gym facilities available to the public. Jiading also has a plan for on-the-water tourism as a promotion of the Moat as a tourist brand. In the future, residents of Jiading will be able to "boat the moat" while enjoying the beautiful vision of dialoguing with history.

Localized Site Project in Jiading

The site project "Create a New Waterfront Life Together" consists of four thematic exhibition – contemporary art exhibition "River Map", Exhibition of Excellent Street Furniture/Facilities of Jiading Moat, Exhibition of Jiading Urban and Rural Planning, and image exhibition "Yesterday, Today and Tomorrow of Jiading Moat" – and three serial activities, that is, a high-end forum, an orienteering around Moat Promenade and Call for Designs of Jiading Moat Street Furniture.

Contemporary Art Exhibition "River Map"

The contemporary art exhibition "River Map" is created by seven active young contemporary artists: Li Fubiao, Li Xianyang, Lin Sen, Pang Hailong, Yu Yang, Zhu Yong and Zheng Xiaohui. They have contributed more than ten of their recent art pieces and installed these varied works among the green belt of moat promenade, creating a spatial effect and experience closer to "life-scenarios".

This exhibition, focusing on the harmonies of human and nature, offers refreshingly novel experiences and sensations with various artistic media, and forms amongst urban environments an atmosphere where traditional and contemporary arts are engaged in a dialogue and cultural ecologies of urban spaces are activated from a unique perspective. Whey citizens walk into these waterfront spaces, they are entering a colourful and enriched journey of cultures which evokes deep enlightenments within.

Exhibition of Excellent Street Furniture/Facilities of Jiading Moat

The Exhibition of Excellent Street Furniture/Facilities is based on the Call for Designs of Jiading Moat Street Furniture, and invites a number of young designers to redesign the street furniture and facilities of Jiading, the theme being the town's cultural characteristics, from the perspective of public art. Their work will be displayed in the form of panels and installments.

Exhibition of Urban and Rural Planning in Jiading

Based on Jiading District Master Plan (2017 – 2035), the Exhibition of Urban and Rural Planning in Jiading presents a comprehensive vision of urban and rural spaces in the future of Jiading: by 2035, Jiading will be a new-type modernized city as part of Shanghai Metropolitan Region, featuring automobile industry, technology, moral culture, health and scenery. It

嘉定城乡规划展
Exhibition of urban and rural planning in Jiading

充分了解嘉定环城河的昨天、今天和明天。

高端论坛

案例展论坛邀请国内知名的城市规划与设计和公共艺术领域的专家进行经验交流。主持人由同济大学建筑与城市规划学院教授、同济城市规划设计研究院总规划师、韩天衡美术馆总建筑师童明担任。

论坛嘉宾邀请熟悉嘉定的专家共同参与，包括嘉定区博物馆副研究员、上海作家协会会员、上海历史学会会员、嘉定地方史专家陶继明；嘉定市民代表，具有 20 年海上远洋资历的嘉定环城河水运专家陈新宏，共同探讨嘉定地方人文和文化发展。陈新宏船长从水上旅游角度，分享了对环城河"三个空间"发展前景的想法：岸线空间——嘉定环城河步道景观提升；陆上空间——打造西大街风貌区、复建西城门楼，使之与城河的呼应更加紧密；水域空间——水上旅游，可以加以进一步开发。陈船长认为市民、建筑、艺术人文、景观设施、娱乐健身空间等元素并存，才能使申城"千年清明上河图"得以完整呈现。

环城河步道定向打卡赛

环城河步道定向打卡赛选取嘉定环城河两侧城市公共空间，依托定向打卡赛，让市民深度体验环城河滨水及腹地空间，感受滨水空间更新改造的建设成就。

100 位选手分成多路，沿环城河步道奔赴各

also demonstrates the major accomplishments as Jiading comprehensively carries out plans of its country units, or rural towns, in close connection to the goals of "rural revitalization" proposed by the Jiading District Committee of Communist Party of China and Jiading District Government.

Image Exhibition "Yesterday, Today and Tomorrow of Jiading Moat"

The image exhibition "Yesterday, Today and Tomorrow of Jiading Moat" at "Now Factory" creative office park (70 Bole Road) and public waterfronts of Jiading Moat, is based on *The Story of Jiading* edited by Gan Yongkang and aims to allow citizens to fully understand the yesterday, today and tomorrow of Jiading Moat in the form of photographs.

High-end Forum

The forum invites a number of nationally renowned professionals in urban planning/design and public arts to share their insights and experiences. The event is hosted by Tong Ming, professor of College of Architecture and Urban Planning Tongji University, chief planner of Shanghai Tongji Urban Planning & Design Institute and chief architect of Han Tianheng Art Museum.

Invited to the forum to discuss the local cultural development of Jiading are several professionals familiar with the town of Jiading, including Tao Jiming, associate researcher of Jiading District Museum, member of Shanghai Writers Association and Shanghai Historical Society, expert in the history of Jiading; and Chen Xinhong, citizen representative of Jiading, expert in water transport of Jiading Moat and seaman of 20 years' experience. Captain Chen, from the perspective of on-the-water tourism, shares his insights about how to develop the Moat's "three spaces": waterfront, by improving the sceneries of Moat Promenade; hinterland, by developing the scenic zone of West Main Street, rebuilding the west town gate and making closer connections to the moat; and waterbody, by developing on-

个点位，完成指定任务。打卡点包含了南水关、古城墙、法华塔、孔庙等多个嘉定文化地标，能让市民在深度体验环城河滨水及其腹地空间的同时，感受城市的发展魅力和深厚的文化底蕴。参赛选手表示，沿途无论是两岸风光、绿植绿化，还是不同主题的现代彩绘及南水关、老城墙等反映嘉定历史的遗迹，都让他们体验颇丰。

环城河步道城市家具设计方案征集评选活动

嘉定环城河步道城市家具设计方案征集评选活动围绕滨水主题和嘉定特色，面向艺术家及公众进行城市家具的作品征集，作品经初选后将在嘉定实践案例展活动当天进行现场

the-water tourism. Captain believes that a "millennial Shanghai version of Along the River During the Qingming Festival" is only complete with a palette of elements such as civil, architecture, art, landscape, entertainment and gym.

Moat Promenade Orienteering

Urban public spaces alongside Jiading Moat are selected as the course of Moat Promenade Orienteering, aiming to involve citizens for a deep experience of the waterfronts and hinterlands of the moat, as well as the accomplishments of waterfront renovation and transformation.

100 orienteers are assigned to several routes, trying to accomplish their tasks by reaching the check points which cover a number of local cultural landmarks, such as the south water pass, the ancient town walls, Fahua Pagoda and Confucian Temple. Amidst the thorough experience of waterfronts and hinterlands of the moat, they are able to feel the developmental

环城河步道定向打卡赛
Moat promenade orienteering

投票，并在上海城市空间艺术季、上海嘉定等公众号平台上进行线上公开投票，以网络投票及活动现场投票相结合的形式，评定产生1名一等奖、2名二等奖、3名三等奖和10名佳作奖。

接轨古城保护规划更新，口碑、热度全面领跑

2019年11月30日，为期2个月的"同创水岸新生活"实践案例展圆满落幕。展览通过充分讲述环城河的新老故事，展示规划、建设和更新提升的工作历程和工作成效，让公众深度参与滨水区活动，并从专业角度探讨推进城市空间艺术品质以及城市精细化管理的路径和机制，引发了社会各界对城市空间品质提升的思考。

笔者从事上海城市空间艺术季活动的推进执行工作，是生活居住在嘉定古城的新上海人。在持续跟进嘉定区实践案例展的工作中，对嘉定古城一朝一夕的城市变化的感受和理解也逐渐加深，也更喜爱上这座古城。

上海城市空间艺术季强调"空间"，在空间

potential and cultural heritages of Jiading. The orienteerers state they are treated with an enriched experience by waterfront sceneries, vegetation, contemporary paintings of various themes and historical relics such as the south water pass and the ancient town wall.

Call for Designs of Moat Promenade Street Furniture

Call for Designs of Moat Promenade Street Furniture is an event calling the artists and the public for designs of waterfront-themed street furniture which reflect characteristics of Jiading. The candidates, after the preliminary screening, will be voted both at the venue of Jiading site project and WeChat subscription accounts of SUSAS and Shanghai Jiading. 1 first-class award, 2 second-class awards, 3 third-class awards and 10 recognition awards will be selected based on the results of on-site and online votes.

Reputed and Popular Site Project in Alignment with Protection, Planning and Regeneration of Jiading Old Town

The successful 2-month "Create a New Waterfront Life Together" Site Project was closed at 30 November 2019. By thoroughly telling the old and new stories of Jiading Moat, demonstrating how and what planning, construction and regeneration efforts have improved the area, involving the public to experience waterfronts, and discussing the approaches and mechanisms

嘉定环城河步道城市家具设计方案征集评选活动（图片来源："上海嘉定"公众号，2019年10月18日）
Call for Designs of Jiading Moat Promenade Street Furniture (source: WeChat subscription account "Shanghai Jiading", 18 October 2019)

上做艺术、建筑、景观等的加减法。实践案例展重点在于案例的在地性展示，其策划、执行、艺术品布展等必然会受到时间和空间的限制。不同的空间和时间条件也会孕育、碰撞出不同的策展和布展模式。在地案例展示的主办方可以考虑邀请艺术家设计师根据展览主题要求、场地情况来进行作品创作，但若面临资金和时间不很充沛的情况，在固定的展期时间内，选择邀请现有的、与主题相关的作品进行临时展示，可以作为艺术介入公共空间的一种可操作的、可持续的、灵活的展览模式。户外空间进行相关主题的艺术展览，能很好地激发出此地的空间活力和烘托当地文化氛围，具有创意的公共艺术作品也能更多地吸引当地艺术爱好者，激发关于家乡、身份、社区变化和社群等议题的温馨讨论。

临时的户外展览相对于固定场馆内的展览来说，是一种从室内"走出来"，更加面向开放、面向公共的展览模式，对策展方和执行方的挑战也更大。展览空间是完全开放性、纯天然的户外环境，艺术品会直接受到气候变化、安全因素、用水用电等因素的影响。展览的公共性又要求策展人要对当地空间、历史、当地文化、建筑、景观等特别敏感，有自己的特殊理解和跨界思维，能善用空间特点，并能引导艺术家根据空间特点和在地人文特点来选择或创作作品。

本届案例展以艺术展览和文化艺术活动作为桥梁，一端连接传统，一端连接当代，使得城市文化遗产与当代艺术在古城河道的贯通空间上产生共鸣和碰撞，给这个底蕴深厚、历史悠久的地方营造一些新的沉淀和空间，让市民从微观与宏观、人文与地理的多角度，感受家乡的过去、现在和未来。

to promote artistic qualities of urban spaces and delicacy management of cities from the perspective of professionals, the exhibitions have aroused various elements of the society to think about how to improve urban spaces.

I am a member of the execution team of SUSAS, and also a new citizen of Shanghai settled in the ancient town of Jiading. As I follow through the Jiading site project, I have ever more experienced and understood this ever changing historical town, and falling more in love with it.

SUSAS, emphasizing the concept of "space", has been adding to or simplifying spaces in terms of art, architecture and landscape. Site projects feature locality, and thus their curation, execution and arrangement must face restrictions in both time and space. Different spatial and temporal conditions will also interact and foster a variety of curatorial patterns and arrangements. The host of a site project may invite artists and designers to create works based on the theme and site conditions, but another plausible, sustainable and flexible approach to introducing art into public spaces, in case of limited funds or constrained time, is to exhibit related existing artworks for a definite period of time. Outdoor and locally themed art exhibitions may serve to energize the venue and promote the neighborhood culturally, and creative public art can be a magnet for local art enthusiasts and arouse warm discussions about changes of hometown, identities and communities.

In contrast to exhibitions in permanent institutions, temporal outdoor exhibitions mark a more open, public and "going-out" model which is also more challenging for both curators and executors. Since the exhibition is totally exposed to the open natural environments, the exhibited items are directly influenced by weather, security, water and electricity supplies. And the public nature of such exhibitions means that curators must be extremely sensitive to local spaces, histories, cultures, buildings and landscapes, have uniquely interdisciplinary understandings and a good command of characteristics of the venue, and guide the artists to select or create works according to features of the site and local cultures.

The Jiading site project constitutes a bridge of art exhibitions and cultural events that connects traditions to our time, and triggers cultural relics of the town and contemporary art to echo and collide in the connected spaces of ancient moat. While it has created new heritages and spaces for the historical town, it also allows its citizens to feel the past, present and future of hometown from multiple perspectives – macro and micro, humanity and geography.

青浦环城水系公园空间艺术实践案例展

Site Project of Qingpu Round-city Water System Park

施皓
华东建筑设计研究院有限公司

季永兴
上海水利工程设计研究院有限公司

夏健
上海淀山湖新城发展有限公司

Shi Hao
East China Architectural and Design Institute

Ji Yongxing
Shanghai Water Engineering Design & Research Institute

Xia Jian
Shanghai Lake Dianshan Newtown Development Co.,Ltd.

主办单位：
上海市青浦区人民政府
承办单位：
上海淀山湖新城发展有限公司
协办单位：
青浦区规划和自然资源局
策展人：
施皓
展览时间：
2019 年 10 月 15 日—2020 年 1 月
展览地点：
青浦环城水系公园水城门、皮划艇仓库、
沿河步道、梦蝶岛

Host:
People's Government of Qingpu District, Shanghai
Organizer:
Shanghai Lake Dianshan Newtown Development Co., Ltd.
Co-organizer:
Qingpu District Planning and Natural Resources Bureau
Curator:
Shi Hao
Time:
15 October 2019 – January 2020
Venues:
Water Gate, Kayak Warehouse, Riverside Promenade, Butterfly Dream Island in Qingpu Round-city Water System Park

2019 年 10 月 15 日，一场盛大的灯光秀表演，拉开了 2019 年上海城市空间艺术季青浦环城水系公园空间艺术实践案例展的序幕。绚丽的投影将迷人的青浦风情投射在刚刚落成的青浦漕港水城门桥的城墙上，音乐时而悠扬，时而激越，展示着青浦这个江南水城在新时期的城市更新中所焕发出的迷人风采。

青浦属于典型的江南水乡，是上海水文化和古文化的发祥地。它既是上海"水乡之源"，境内有 93 个湖泊和 1934 条河道；又是上海"母亲河之源"，苏州河、黄浦江均发源于此；也是上海"饮用水之源"，有黄浦江上游水源地和金泽水库；更是上海"文化之源"，有历史悠久的崧泽文化、福泉山文化等遗址。因此，水作为青浦的特色资源和城市名片，在青浦经济社会发展中具有突出的地位和重要的作用。

位于青浦核心地块的青浦新城有 31 条河流水系。青浦新城的核心区有约 30 万居民，

A magnificent light show at 15 October 2019 marked the opening of the SUSAS 2019 site project of Qingpu Round-city Water System Park. The charms of Qingpu were marvelously projected onto the recently completed Qingpu Water Gate on Caogang River while the music, once melodious and once exciting, demonstrated how the Jiangnan water town of Qingpu was becoming enchanting in this new ear of urban regeneration.

Qingpu is a typical Jiangnan water town, and is where the rivers and civilization of Shanghai start. It boasts of plenty waterbodies, including 93 lakes and 1,934 waterways. It contains the sources of Shanghai's "mother rivers", Huangpu River and Suzhou Creek. It provides water which Shanghai citizens drink, with its upper Huangpu headwater and Jinze Reservoir. And it is the cradle of civilization in Shanghai – you can find Songze Culture and Fuquan Hill Culture. Therefore, water, as a unique asset and signboard of Qingpu, plays a prominent and essential role in the social and economic development of the town.

31 rivers flow through Qingpu New City at the centre of Qingpu District. While the core of New City is populated by about 300,000 people, its 31 rivers have yet to be functional and serve its residents. As the recent rapid industrialization process massively transforms the appearances of the traditional water town, old-style streets, lanes, homes and scenic rivers are slowly

青浦漕港水城门桥开幕灯光秀
Opening light show at Qingpu Water Gate on Caogang River

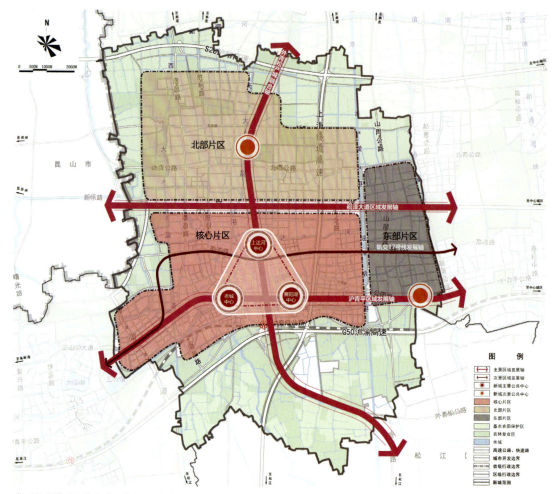

"一城两翼"发展格局（图片来源：青浦区规划和自然资源局）
"One City and Two Wings" Strategy (source: Qingpu District Planning and Natural Resources Bureau)

但 31 条河流水系尚未发挥其应有功能，为居民提供更多服务。近年来由于城镇工业化建设高速发展，其传统的水乡风貌发生了巨大的改变，水乡街巷、传统民居、风貌河道正在慢慢消逝。

随着"一城两翼"的青浦区总体发展战略格局的提出，青浦新城，作为发展引擎之一，规划将其建设成为长三角地区综合性节点城市。发展的使命需要青浦新城提升城市品质和竞争力，现代生活也对城市环境和空间提出了更高要求。青浦新城应该有温度、有记忆、可漫步、可阅读。

因此，青浦区将目光放在了城中纵横交错的河道，将封闭已久的滨河水岸重新开放给市民，通过对环青浦主城区的四条大河的滨河生态景观的营造，运用滨水生态修复和"还岸于民"空间更新相结合的设计理念，以水

vanishing.

According to the proposal of Qingpu general development strategy, "One City and Two Wings", Qingpu New City is a growth engine of the district and is planned to be a comprehensive node in the Yangtze River Delta region. Its developmental mission means that Qingpu New City must increase its quality and competitiveness, and modern lifestyles have also imposed higher demands on urban environments and spaces. Qingpu New City should be a warm, memorable, wanderable and worth-reading town.

So Qingpu District has focused on its crisscrossing network of rivers, trying to make the long closed waterfronts again available to the citizens by developing the ecological waterfront landscapes of the four major rivers flowing through downtown. Under the concepts of restoring waterfront ecologies and renovating spaces to "return the banks to the people", the basic measures are water engineering projects which restore the aquatic ecologies, transform the ecological landscape on bank protections and rectify and connect flood prevention courses, while additional functions of wetland, landscape, culture, leisure

青浦环城水系公园（图片来源：青浦区规划和自然资源局）
Qingpu Round-city Water System Park (source: Qingpu District Planning and Natural Resources Bureau)

利工程的河道水系生态修复、护岸生态景观改造、防汛通道贯通整治为基础，融入湿地、景观、文化、休闲、运动等其他功能元素；通过环城水系整治、滨水空间开发和重要节点建设等技术措施，打造集"防洪排涝、生态景观、文化旅游、休闲娱乐、城市形象"于一体的青浦环城水系公园，重塑上海水城，找回青浦乡愁，使青浦新城区成为具有"水乡文化"和"历史文化"内涵的生态宜居城区。

在具体的设计实施中，根据青浦新城总体规划，对环城水系内的所有河道两岸的用地性质和功能均进行了明确。位于新城北部的上达河，规划北岸为产业区，南岸为商业区，河道中部规划建设商业街区，地标性建筑"青浦中心"和地铁17号线"青浦新城站"均位于此，该段由此以滨水商业功能植入。东部的油墩港因两岸地带宽阔，且有架空高压线限制，适宜建设湿地和休闲活动，所以以生态观光功能打造。南部的淀浦河因有崧泽遗址、老城厢、万寿塔等历史文化遗迹，并新建了南菁园、浦仓路人行桥等特色建筑，所以该段定位为文化体验段。城区西部的西大盈港因两侧为大量居民小区，对休闲、运动的功能要求多，所以该段以植入运动休闲功能为主。

and sports are involved. The Qingpu Round-city Water System Park, combining elements of "flood prevention, ecological landscape, cultural tourism, leisure and city branding", is created by renovating water systems around the city, developing waterfront spaces and constructing key nodes. These measures aim to reshape the water town of Shanghai, bring back the memories of Qingpu and make Qingpu New City an ecological and habitable urban area featuring "Water Culture" and "Historical Culture".

When it comes to implementation, all lands along both banks of involved rivers are assigned definite usages and functions according to the master plan of Qingpu New City. The north bank of Shangda River at the northern part of Qingpu New City is intended to be an industrial area, its south bank a business hub and its central section a retail waterfront where the landmark of "Qingpu Centre" and the "Qingpu New City" metro station of Line 17 stand. The eastern Youdun Stream, for its broad spaces on either bank and overhead high-voltage wires, is suitable for an ecological sightseeing area featuring wetlands and leisure. In the south is Dianpu River around which are, in addition to historical heritages such as the relics of Songze Culture, traditional Shanghainese streets and Longevity Pagoda, Nanjing Garden, Pucang Road Footbridge and other new characteristic structures, so the river is designated for cultural experiences. The West Daying Stream, the western quarter of Qingpu town, is surrounded by many residential neighborhoods in high demands for leisure and gym, so satisfying these demands are the primary concern in this area.

To alleviate the problems that rivers in Qingpu tend to be straight

另外，为改善打破青浦河道平直、线型的缺点，结合几个交叉河口，拓宽水面，形成几个较大水面的湖区。在上达河与东大盈港交叉口，拓宽水面，形成智慧湖公园；上达河、老西大盈港、东大盈港围合，形成了一个狭长岛屿，因此规划建设一座长岛公园；在淀浦河与漕港河交叉形成的半岛，打造大型活动场所——梦蝶岛公园；在油墩港和淀浦河交叉口，利用东岸的崧泽遗址博物馆和高压线下空地拓宽成的湿地共同打造崧泽遗址公园。另外，利用万寿塔和西大盈港双桥周边的地块，打造古塔公园、双桥公园。这样，通过更新改造形成的六个景观节点与既有的曲水园、夏阳湖等两个景观节点，共同丰富了城区的滨水活动空间。

其在具体的措施上运用了基于亲水生态的既有护岸改造，以及布置码头，开辟水上游线等方式，使人和水的关系更为亲密；在沿水边条件适合位置布置大量湿地，以自然做功的方式，实现对水体的净化，逐步恢复本地区生物多样性；将青浦历史人文在景观中再现，恢复历史上的水城门、青溪书院等景点，并将名人典故，诗词文赋等体现在众多新建的桥梁命名上，让市民体会背后的青浦故事；将休闲体育等活动植入，贯通河道两岸 4 米的步道，形成 21 公里长的半马跑道，结合周边住宅布置体育场地，为城市生活注入活力。

通过环城水系公园的建设，青浦人和水再次亲近，城和水再次相融，岸与绿再次交织。

为体现青浦水城一体的特点，本次青浦实践案例展的策展，在 2019 上海城市空间艺术季的"相遇"这个大主题下，提出了"所谓伊城，在水一方"的主题。通过对于中国人耳熟能详的《秦风·蒹葭》中名句的再演绎，体现了青浦人与水共生，与水共情的生活；青浦城以水建城、以水兴城、以水养城、以水荣城的发展历程；青浦与水割不断，解不开的情愫。

展区分为两处。主展区布置在青浦城区西南角的淀浦河与漕港河交汇处的水城门的周围。这里是青浦老城的所在地，历史上，青浦是

and linear, we have created several considerable lakes by widening river crossings: Wisdom Lake Park at the crossing of Shangda River and East Daying Stream; Long Island Park at the bar surrounded by Shangda River, East and West Daying Streams; the massive Butterfly Dream Park at the peninsula by Dianpu River and Caogang River; and Songze Site Park at the crossing of Youdun Stream and Dianpu River, based on the Songze Site Museum on the east bank and the wetland which used to be an unused space under high-voltage wires. What's more, Pagoda Park and Twin Bridges Park is constructed out of the plots in adjacent to Longevity Pagoda and the two bridges on West Daying Stream. These six new or renovated scenery sites, with the existing Qushui Park and Xiayang Lake, have enriched the waterfront activity zones in the town of Qingpu.

In addition to these basic actions, the bank protections are renovated based on waterfront ecologies, and cruise docks are installed to get humans and waters closer. Plenty of wetlands arranged at suitable locations along the rivers are expected to purify water by the force of nature and gradually restore local biological diversity. The historical Water Gate and Qingxi Academy are restored to represent the histories and cultures of Qingpu in the form of landscape, and anecdotes of renowned figures, classical poems and proses are embodied in many new bridges, informing the citizens of the stories of Qingpu. And the new facilities for physical exercises, including the four-metre-wide promenades on both banks of the developed rivers that constitute a course of half-marathon (21km), as well as the gym spaces around residential blocks, have introduced more energy into city life.

The project of Qingpu Round-city Water System Park has brought the residents of Qingpu and its waters intimate again, integrated the town of Qingpu and its flows again, and weaved the banks and the greens of Qingpu again.

To reflect the water-and-town-as-one nature of Qingpu, the curators of Qingpu site project proposed the theme of "Where's She I Need? Beyond the Stream", adapted from a poetic line of *Book of Songs: The Reeds* widely known in China, under the general theme of SUSAS 2019, "Encounter". The line, originally a call for a beloved fair lady, now represents the life of Qingpu in an empathetic symbiosis with water, the history of Qingpu as it is created, nurtured and glorified by waters, and the inseverable and unbreakable sentimental bonds between Qingpu and water.

The site project is distributed in two venues. The main venue near the water gate which sits at the southwestern corner of Qingpu, the crossing of Dianpu River and Caogang River, is part of the old town of Qingpu. Qingpu is a historically a "town of fish and rice" where ships used to carry the abundant local produces – rice and silk threads - through the dense river networks to Dianshan Lake and Tai Lake, then enter the Grand Canal and finally reach every corner of China. The Water Gate

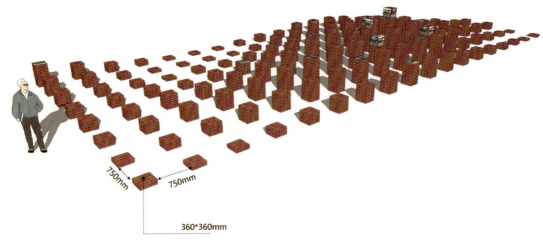

波浪点阵 / 近代（图片来源：施皓）
Wavy dot array / Early Modernity (source: Shi Hao)

照片墙（图片来源：施皓）
Photo wall (source: Shi Hao)

智慧湖公园（图片来源：青浦区规划和自然资源局）
Wisdom Lake Park (source: Qingpu District Planning and Natural Resources Bureau)

"鱼米之乡"，来来往往的船只运载着青浦丰富的物产，从这里出发，通过纵横的水网，穿过淀山湖、太湖，汇入大运河，将青浦的稻米、蚕丝等运送到全国各地。这里象征着青浦得天独厚的天然禀赋，漕港河成为青浦的经济之河。而淀浦河作为青浦的主要干流，连接了淀山湖和黄浦江，源远流长的古文化在淀浦河沿岸生根发芽，体现着青浦厚重的历史文化底蕴。王昶在这里修志教书、四方乡民在万寿塔下祈求风调雨顺。淀浦河是青浦的文化之河。在环城水系的建设过程中，规划设计充分发掘了这块土地上的历史故事，用一座具有象征意义的水城门步行桥和桥两岸带有新中式风格的开放广场，重新构建和定义了当代的青浦西南大门。水城门步行桥已经成为了新时期青浦城市的地标之一。

此次策展，通过艺术装置的布置，展示新时期青浦人、城、自然的和谐关系，展示环城水系建设之后，青浦欣欣向荣的社会景象。其具体体现在三个"四"：展示青浦人的四方笑容，展示青浦城的四时变迁，展示青浦自然的四季美景。

其在广场上，以四组不同材质的波浪形点阵，体现青浦城市的四时变迁。藤编代表着古代，红砖代表着近代，环保材料代表着现代，而

is a symbol of the brilliant natural endowments of Qingpu, while Caogang River is its economic artery. Dianpu River, as the primary river of Qingpu, connects Dianshan Lake and Huangpu River, nurtures ancient civilizations flourishing on its banks and thus embodies the historical and cultural treasures of Qingpu. Wang Chang compiled and taught here. The Longevity Pagoda attracted villagers from the entire region to pray for good climate. Dianpu River is the cultural river of Qingpu. As the water system is developed, histories and stories of Qingpu is thoroughly incorporated into urban planning and design. An iconic water gate which also functions as a foot bridge and the neo-Chinese-style open square on its banks have restructured and redefined a contemporary southwestern access of Qingpu. The Water Gate has already become a new landmark of Qingpu.

The site project, in the form of installment art, demonstrates the harmonious relationship among human, city and nature in the new era of Qingpu, and the prosperity of Qingpu after the completion of city water system. More specifically, it shows the welcoming smile of Qingpu people, the historical changes of Qingpu town and the beautiful natural sceneries all year around.

Four wavy arrays of nodes on the square represent the four eras of Qingpu: rattan for the ancient period, red bricks for early modernity, environment-friendly materials for the present, and transparent glass bricks for the future. While the materials are changing, the wavy pattern symbolizes the constant entanglement between water and Qingpu. Each node is topped with a picture about the culture of Qingpu.

Off the water gate is a massive gallery frame where two photo walls, as a tribute to the happy life and natural beauty of Qingpu,

透明玻璃砖代表着未来，变化的是材料，点阵顶面的波浪起伏，象征着水与青浦不变的缠绵交织。每个点阵的端头都用照片展示着青浦的风土人情。

而水城门不远处岸边的大廊架中，布置了两块大的照片墙，通过对青浦市民笑容的采集以及青浦四时美景的图像展示，体现着青浦人的生活之乐、自然之美。在水城门上的城楼里，布置了对于环城水系公园以及青浦区的介绍，以设计图纸和宣传视频滚动播放的形式，介绍了新时期青浦的发展以及对美好未来的展望。

另一处展区，位于上达河和东大盈港交汇处的智慧湖公园。这里是青浦新城规划中未来发展的核心。公园规划了一处水广场，以震撼富有动感活力的水幕电影，在夜晚将青浦人汇聚到这里，为他们提供一处交流、休憩、聚会的场所，展示着青浦平时波澜不兴之下涌动的时尚和活力。广场一旁的电子屏幕和主展区一样，滚动播放着青浦的宣传视频，形成展示上的联动，相得益彰。

优越的生态环境，浓郁的水乡特色，深厚的文化底蕴是青浦由传统工业文明转型绿色经济发展的重要驱动内核。青浦环城水系经"防洪、生态、景观、历史、人文"五位一体的水生态修复与滨水空间提升改造，形成了青浦新城新的"一环、四段、八园"的空间总体布局，丰富了青浦"水乡文化"和"历史文化"，重塑了上海水城，带动了青浦环城滨水区域的产业升级，通过释放滨水空间，让青浦市民和游客可以共享绿水美景，充分发挥城市水系"防洪除涝、生态景观、运动休闲、城市环境"的综合功能，助力青浦打造生态宜居新城。

人离不开水，"仁者乐山，智者乐水"；水赋予了城市灵魂，"近山则诚，近水则灵"。人和城的相遇也许是偶然，但人和水的相遇是必然。或许有一段时间，人水曾相隔，溯洄从之，道阻且长，但是兜兜转转当再次相遇，必能迸发新的精彩。

show the smiling faces of Qingpu residents and the sceneries of Qingpu in each of the four seasons. Introductions to Qingpu Round-city Water System Park and Qingpu District, in the form of design drawings and looping promotional videos, are made on gate towers as a showcase for the development of Qingpu in this new era and an envisioning for the promising future of Qingpu.

Another exhibition venue is the Wisdom Lake Museum at the crossing of Shangda River and East Daying Stream. The planning of Qingpu New City has designated this area as the heart of its future development. An aquatic square is planned in the park where shockingly dynamic water curtain movies will act as a nocturnal anchor for citizens to meet, rest and gather, demonstrating the fashion and vitality beneath the tranquility of Qingpu in its daily life. The screens by the square, like those in the main venue, loop promotional videos of Qingpu and thus form an interactive echoing effect.

The essential motivating engines of Qingpu, as it transforms itself from conventional Industrialism to green economy, are its excellent natural environment, its rich flavors of water-town and its treasure of cultural heritages. After the comprehensive ecological restoration and waterfront improvement project of Qingpu Round-city Water System that encompasses five aspects – flood prevention, ecology, landscape, history, culture – Qingpu New City has achieved a new overall spatial pattern, that is, "One Ring, Four Quarters and Eight Parks". The "water-town" and historical cultures of Qingpu are enriched, the water town of Shanghai is reshaped, and the industries around the Qingpu Round-city Water System have upgraded. By opening the waterfronts, residents of and tourists to Qingpu can share the wonderful natural landscapes. Thanks to the project, the water systems of Qingpu can fully realize their functions of flood prevention, natural landscape, leisure and gym amenities and urban environments, helping Qingpu to create an ecological and habitable new city.

Humans cannot live without water – "the knowledgeable love waters, the benevolent love mountains". Water fills cities with souls – "living at mountains makes one honest while living by waters makes one enlightened." It might be an occasion that a person meets a city, but it's necessary for every person to encounter water. A person might be separated from waters for some time, and a long and difficult way lies between them. But once they meet each other again after much detouring and revolving, new sparkles would definitely be ignited.

乐水——长宁区苏州河实践案例展

The Water of Promise — Site Project at Suzhou Creek Waterfront, Changning District

马宏
上海城市公共空间设计促进中心
同济大学博士研究生

Ma Hong
Shanghai Design and Promotion Centre for Urban Public Space
Ph.D Candidate in Tongji University

主办单位：
上海市长宁区人民政府
承办单位：
长宁区规划和自然资源局、长宁区建设和管理委员会、长宁区绿化市容局、周家桥街道
协办单位：
长宁区体育局、长宁区文化和旅游局
策展人：
李丹锋、周渐佳
展览时间：
2019 年 10 月 25 日—2019 年 12 月 6 日
展览地点：
长宁区长宁路天原河滨花园（长宁路芙蓉江路至天中路）

Host:
People's Government of Changning District, Shanghai
Organizers:
Changning District Planning and Natural Resources Bureau, Construction and Traffic Committee of Changning District, Administration of Greening and City Appearance of Changning District, Zhoujiaqiao Sub-district Office
Co-organizers:
Changning Administration of Sports, Changning District Bureau of Culture and Tourism
Curators:
Li Danfeng, Zhou Jianjia
Time:
25 October 2019 – 6 December 2019
Venue:
Tianyuan Riverfront Park, Changning Road, Changning District (surrounded by Changning Road, Furongjiang Road and Tianzhong Road)

2019 长宁区实践案例展围绕一江一河贯通工作，回应滨水生活主题方向，将苏州河景观带长宁路段作为参展项目。建成近 10 年的苏州河景观带长宁路段，东起虹桥河滨花园，西至哈密路，滨河岸线长 2.5 公里，其中 2.1 公里长宁路紧贴苏州河。在上海市重点打造"一江一河"滨水生活岸线的今天，此段滨水景观带仍然受到众多好评。苏州河景观带长宁路段既属苏州河沿岸与绿地网络，又在慢行网络之中，绿地北面是苏州河滨水步行道，南面是包括仁恒河滨花园、天山河畔花园在内的居住区，既服务于社区居民，同时也向城市开放。始于十年前的建设，既有前瞻性，也通过多年的不断调整、维护增加了丰富性，具备了探讨、实验的价值。

2019 年，长宁区积极响应市委市政府"一江一河"贯通开放的要求，11.2 公里的苏州河岸线在全市率先实现贯通。新的贯通工程更加重视人性化，加强实用性，体现精致独特性。在这一新的发展背景下，将十年前建成的项

The SUSAS 2019 site project of Changning District is centred around the connection project of Suzhou Creek and Huangpu River, and the Changning Road section of Suzhou Creek Scenic Belt is selected as the venue, echoing the theme of waterfront life. This section, completed for nearly ten years, stretches eastward from Hongqiao Waterfront Park to Hami Road and of its 2.5km waterfront, 2.1km of Changning Road are closely adjacent to Suzhou Creek. While Shanghai has been focusing on developing lifestyle waterfronts along Suzhou Creek and Huangpu River, this section is still well reputed. The Changning Road section of Suzhou Creek Scenic Belt is part of Suzhou Creek Waterside Green Belt and a slow-traffic network, lies between the Suzhou Creek waterfront promenade at the north and residential neighborhoods - Hengren Riverfront, Tianshan Riverside and more – at the north, and thus both serves residents and opens to the entire city. As a project initiated ten years ago, it is both farsighted and worth discussion or experimentation for all the richness bestowed by years of adaption and maintenance.

In response to the demands of the Party Committee and Municipal Government of Shanghai to connect and open Suzhou Creek and Huangpu River, Changning became the first district which finished the connection project (11.2km of Suzhou Creek waterfront) in 2019. The new connection project focuses

居民的休闲活动
Leisure activities of residents

居民的休闲活动
Leisure activities of residents

目参与空间艺术季，让市民了解曾经的建设成果及建设历程，思考如何不断贴合当代市民的需求更新发展，满足市民对美好生活的期盼是此次案例展更重要的社会意义。

本次案例展选址于景观带中的天原河滨花园。花园原为上海天原化工厂厂区，工厂搬迁后，整个基地统一规划建设。早在1999年6月关于该地区控制性详细规划中，就确定了"城市公共绿地沿长宁路布置，有利于增加苏州河沿线绿地面积，提高苏州河沿岸滨河环境景观质量"的目标，此后又先后提出了"应按规划要求与住宅等主体工程同步实施、同步建设"，"滨河绿地建成后应作为公共绿地确保向社会开放"等一系列要求。在此过程中，"为周边居民服务""向社会开放"的定位被始终贯彻与延续。如果仅有规划，而没有好的实施也不会有现在的状态。2008年，长宁区建交委开始组织编制苏州河沿线景观提升改造设计，并将其作为"迎世博"的重点工程。通过多家方案比选，澳大利亚BAU建筑与设计事务所的方案胜出，2009年2月设计方案基本定稿后开始实施，并于2010年改造完工。在后续几次养护中，较好地延续了原设计思路，并结合使用和管理在细节上进行了完善，这才使景观带历经十年依旧维护良好。

本次案例展主要通过三个板块，即"装置搭建""案例展示"和"研究表现"阐释生活场景如何被创造、滨水空间如何被塑造和在地性如何被营造。三个板块与现有场地条件形成互动，利用场地较长的线性流线，通过特定的展览形式让观者对场所的印象对应。例如出现在绿地里的装置、出现在避雨亭里的研究等，以此加强对空间的印象和场地的认知，也通过设置室内外的导视系统引导流线。

通过各种别具匠心的公共艺术品、城市家具散落此段滨江绿地中，在花园蜿蜒曲折的空中走廊上，则主要展示实践案例、历史研究等板块。围绕展览场地讨论伴随苏州河而生长的城市记忆、生活方式和各种体验。展览期间组织了包括苏州河慢行、公共论坛等活动，进一步丰富关于滨水生活的讨论与畅想。

more on human experience and functionality to reflect unique delicacies. Against this new developmental background, it is even more socially significant to present a project completed a decade ago to SUSAS so that citizens may learn about what and how these construction works are achieved, and think about how developments may keep adapting to the contemporary needs of the public and fulfilling the citizens' hopes for lovely life.

The site project is located at Tianyuan Riverfront Park as part of Suzhou Creek scenic belt. The site of Tianyuan Riverfront Park, formerly Tianyuan Chemical Plant, was planned and constructed as a whole after the plant moved out. The detailed controlling plan of the site, dated June 1999, already decides on the goals that "urban public greens should be arranged along Changning Road to increase greens and improve environmental sceneries along Suzhou Creek". Later, a series of demands were made such as "[greens] should be implemented and constructed simultaneously with the main project of residential blocks" and "should be open to the public." In this process, the greens of Tianyuan Riverfront Park have continued and maintained the positioning to "serve citizens in the neighborhood" and "open to the public". Planning alone, if without proper implementation, would not achieve the results as they are now. In 2008, Changning Construction and Transportation Committee initiated the documentation of landscape improvement project along Suzhou Creek, and enlisted it as a key project to welcome EXPO 2010. After comparing designs proposed by multiple agencies, Brearley Architects and Urbanists (BAU) from Australia won the bid. The design was finalized in February 2009 and the renovation project was completed in 2010. Several rounds of maintenance have successfully kept in line with the design intent and improved on a number of details based on experiences of users and administrators, and that's how this project is still running well after ten years.

The site project consists of three sections – Installment, Case Story and Research – in order to explain how life scenarios are created, how waterfront spaces are shaped and how locality is fostered. The three sections, interactive with existing conditions, take advantage of the long and linear site to match specific forms of exhibition with its impressions upon visitors. For instance, installments on lawn and a research project in a rain pavilion are intended to leave stronger impressions and perceptions of the site, and signage and wayfinding systems, indoor and outdoor, are placed to guide pedestrian flows.

Uniquely crafted public artworks and street furniture distributed among waterfront greens and the zigzagging air passageways in the garden focus on case stories and historical studies, and offer discussions about the city memories, lifestyles and various experiences that grow with Suzhou Creek. Furthermore, a series of events such as Wander along Suzhou Creek and Public Forum are held during the period of exhibition, making more contributions to discussing and envisioning our waterfront life.

天桥
Footbridge

策展人说

策展人李丹锋、周渐佳在采访中说道，城市更新是空间艺术季的大背景，但是 2019 年更多地聚焦在"一江一河"滨水空间的品质提升、为民所享上，主展场选在杨浦滨江，长宁区案例展选在了沿苏州河的天原河滨公园，形成了很有意思的呼应。

苏州河贯通是长宁区近几年在推动的重要工作。在一开始接到这个任务时，其实长宁区规划资源局、建设交通委曾提供了好几个备选的点位，包括临空滑板公园、虹桥河滨休闲公园等，最后结合场地条件，还是一起定了天原河滨花园。与案例展同步，苏州河长宁段刚刚完成了华政段的贯通，尽管尺度小，但是难度不亚于任何一个新建项目。从2020 年来看，苏州河长宁段几近全线贯通，沿着苏州河也形成了一条滨河慢行线，把沿苏州河的所有公共空间串联起来，这样看起来也更理解当时这些案例展选点对于整个计划的意义。当然，我们很高兴以一种特别的方式参与到整个苏州河长宁段贯通的进程中，展览中也以更加接近生活化的方式向大众介绍了这些项目。

Curatorial Comments

The curators, Li Danfeng and Zhou Jianjia state in interviews that while urban regeneration has been the big picture of SUSAS, SUSAS 2019 focuses on how the waterfronts of Suzhou Creek and Huangpu River are improved and served to the people. In this sense, the main venue in Yangpu Waterfront and the Changing site project at Tianyuan Waterfront Park, which lies along Suzhou Creek, make interesting echoing.

Suzhou Creek Connection is a key task of Changning District in the last few years. Upon receiving the task, Urban Planning and Natural Resources Bureau and Construction and Transportation Committee of Changning District proposed several candidate sites such as Linkong Skatepark and Hongqiao Waterfront Leisure Park, and Tianyuan Waterfront Park is eventually chosen based on its site conditions. As the site project is proceeding, the ECUPL (East China University of Political Science and Law) section of Changning Suzhou Creek Connection Project, whose unimpressive scale doesn't mean that it is less difficult than any new construction, has just been completed. Now from the perspective of 2020 when the Changning section of Suzhou Creek is nearly completely connected and a waterfront slow-traffic line has joined together all public spaces along Suzhou Creek, it seems easier to understand why the location choices of the site projects are essential to the overall program. And we are certainly delighted to participate in the Changning Suzhou Creek Connection Project in such a particular manner, and to see that this project is presented to the public through accessible exhibitions.

天桥桥下空间的《鞬园》
Eng Garden under the footbridge

瞭望塔的灯光改造《光水相》
Phases of Light and Water, light design for the watchtower

如何认识场地的魅力

现在看天原河滨公园，印象最深的可能是繁茂的绿化，还有形状很特别的瞭望塔和天桥。案例展中有很多的展示形式就利用了这些条件，比如说对瞭望塔的灯光改造《光水相》，沿着天桥逐渐展开的苏州河建设项目介绍，以及口述系列《苏州河边，一个家，一爿厂，一段时光》，还有结合天桥中休息亭的《起立》，结合天桥桥下空间的《鞥园》和《山影重重》。这些都是展示上的心思和技巧，希望展览能和场地更加贴近。

更有意思的是对场地本身历史的挖掘，这也是随着策展工作逐渐深入而发现的。天原河滨公园的前身是民族企业天原化工厂，对于苏州河和上海的历史来说都是很浓重、很有代表性的一笔。2000年，工厂迁出原址，改建为天原河滨公园，也将周边大大小小的绿地聚合起来。虽然场地上已经很难感受到当年化工厂的痕迹，但是厂里的员工、见证了这段历史的人们还在，甚至就住在附近。非常令我们自豪的一点是，以发生在这个场地上的案例展为契机，曾经的天原化工厂的员工参与了《城市中国》组织的口述计划，他们也回来看了展览，参加了开幕式。河滨公园的管理方也提出在展览结束后想把这些口述资料永久地保留在场地上。尽管这个过程在视觉上和空间上显得不那么强烈，但这个过程是展览非常重要的组成部分，或者说它拓展了案例展的意义。案例展可以是一种挖掘和建立全新的连接——它在曾经在此地的人、场地和现在的使用者之间建立了连接。

如何构思这个展览

长宁区苏州河实践案例展非常特别的一点是，它是一个露天的展场，所有的作品分布在整个带状的沿河绿地中，它始终鼓励人们用一种动态的、漫游的方式去加入这个展览。基本上，瞭望塔、小广场、绿地、所有桥下空间，甚至小到步廊的柱子、扶手都被动员起来，成为展览的背景或者作品、行为发生的场地。并且，对材料、做法的选择都可以容纳这种露天的展示。我们在策展方案汇报时就有评审专家提出，

How to Understand the Charms of the Site

What's most impressive about Tianyuan Riverfront Park is probably its dense vegetation, the peculiarly shaped watchtower or the Footbridge which are utilized in various ways in the site project: the light design for the watchtower *Phases of Light and Water*, introduction to Suzhou Creek developments that gradually unrolls on the Footbridge, the oratory history *By Suzhou Creek: Home, Factory and Time*, the installment in the pavilion on the Footbridge *Stand-up*, and the art pieces *Eng Garden* and *Mountain Shades* under the bridge. All these embody inspirations and skills which try to bring the exhibition closer to the site.

What's more interesting is the revelation of the site's histories which are only gradually shown with the curatorial process. Tianyuan Riverfront Park used to be Tianyuan Chemical Plant, a modern factory owned and operated by Chinese people in the first half of 20th century and an important, representative element in the history of Suzhou Creek and Shanghai. After its relocation in 2000, the original site has become Tianyuan Riverfront Park which also incorporates the adjacent greens, large or small. Though it's hard to believe that a chemical plant used to stand here, former employees and other witnesses to the history of Tianyuan are still alive, and some of them even continue living nearby. We are very proud to see that the former employees who participated in an oratory program of *Urban China*, given the opportunity that the site project is held where they used to work, come back to attend the opening ceremony. The park administrators suggest to permanently keep the oratory documents in this site even after the exhibition is closed. This process is not visually or spatially remarkable, yet still an integral element or extension of the site project – a site project can be a way to reveal and create new connections among the site, its former occupants and its current users.

How the Exhibition is Conceptualized

A particular feature of Changning Suzhou Creek site project is its exposed venue: all its works are distributed among the entire waterfront green belt, encouraging people to enter the exhibition in a dynamic and wandering manner. The watchtower, the square, the lawn, all spaces under the footbridge, and even the columns and handrails of the gallery are mobilized as background, exhibited items or action fields. And materials and approaches are carefully selected to house such an open exhibition. As early as our presentation of the curatorial program, the evaluation board hoped that the exhibition could be fitted nicely into the park, or daily life itself and, though not necessarily a formal display, offer new possibilities during the period of exhibition.

We love all the artworks. We have discussed with the curators

二层步道上的苏州河案例回顾
Review of Suzhou Creek Site Project on the upper promenade

希望展览能非常恰当地嵌入到公园里，也是日常的生活里，它们未必要显得像陈列式的展览，但是会给这段展期内的使用带来新的可能性。

所有的作品我们都很喜欢，一方面是和参展人就方案做了一些细致的沟通，另一方面，当它们真正在场地中实现时，会看到人们会以一种很意外的方式来回应这些作品。比如说演奏者会对着《起立》里的蓝色小人吹萨克斯风，会有小朋友在《山影重重》下捉迷藏，会有年轻女孩来《漫步道》前停留一会，闻闻香味，或者保洁工人在《第X种相遇》坐一坐，休息一会。这些片段都让人动容，也是一个展览能激发出的可能。这也是我们对待这个展览的初衷，能让大家在游历、使用的过程中就把展览看完了，但是展览本身的内容的挖掘还有它的完成度又是经得起推敲的，所以我们在展览尾声的阶段为"乐水"案例展做了一本成果手册，让展览的内容以不同的形式再做进一步的延展与发酵。

如何通过展览表达生活空间

作为一个没有展厅的展览，需要融化在日常

to the very details of the program, and on the other hand, the artworks have aroused unexpected reactions when they are finally installed in the site – before the blue figurine of *Stand-up* are performers playing saxophone, under the *Mountain Shades* are children playing the hide-and-seek, *Promenade* have attracted young girls to stay and smell the fragrances, and *The Xth Encounter* provides a resting place for cleaners. All these moving scenes constitute a possibility of what an exhibition can provoke. And this is exactly our original hope for the exhibition: finish viewing the exhibition while walking around and using it. Yet there is more to dig about the contents and completeness of the exhibition, so we have produced an "accomplishment brochure" about the site project "Water of Promise" towards the closure, as a fermenting extension of the exhibition in alternative forms.

How to Express Life Spaces through Exhibition

The hall-less exhibition needs to be incorporated into daily life scenarios and usages, and this directly diverts us away from the conventional curatorial approach in which we have a more or less definite estimate about who will come or how informed they will be. But once the site is completely opened up, all the scholarly cognitive or discursive structures have to fall apart because it might well be a random jogger or a mother with her kids enter the park, and thus begin their visit to the exhibition – there is no chance for the curators to completely explain the background and situation. So how to establish some sense of structure here? How to convey meanings? We, with

天山初级中学弦乐团开幕式演出
Tianshan Middle School string orchestra at the opening ceremony

生活的场景中和使用中。这也直接导致了我们在策展思路上的转变，以往会为展览设定大致的观众群或者一个大致的认知基础。但是当场地完全开放后，原有的非常学院派的认知体系或者言说结构就必须被打破，因为很可能只是跑步的人，带孩子的人从某个入口进入公园，就这样开始看展览了，并没有这个机会让策展人完整地把前因后果讲一遍。那么在这样的情况下，这个展览的结构如何能成立？它的意义如何被传达？所以我们和长宁区规划资源局在这方面一起做了很多展览流线上、作品形式上的推敲，也得到了参展人的毫无保留的支持。可能这点也和沿杨浦滨江实现的公共艺术作品有些相似，无论是谁、无论是什么情境下来看，这些作品和它们试图传达的公共性都是成立的。也许这也是将来的艺术季案例展甚至主展览可以考虑的方向，这些作品首先是一种和城市相关的空间实践或者行为实践，但是在空间艺术季期间的方式集中呈现，也许这也是从单纯的展示逐渐走向短时甚至永久实践的可能。

Urban Planning and Natural Resources Bureau of Changning District, have devoted much thought to the flows and forms of the exhibition which are unreservedly supported by the participating artists and designers. This might be a shared characteristic with the public artworks at Yangpu Waterfront since their publicness – and their public significance – is always present no matter viewed by whom against what context. Maybe this is a worthy approach of future SUSAS site projects and even the main exhibition. These public artworks are in the first place a spatial or behavioral practice related to the city, only that they are presented in the concentrated form of SUSAS, and it is possible to shift from mere exhibition towards temporal or permanent practical projects.

结语

"为什么上海要做空间艺术季？"长宁区规划和自然资源局副局长苏立琼认为："举办空间艺术季是希望能把这些优秀的成果拿出来做一些展览展示和公共活动，让老百姓更加理解什么是值得倡导的规划和设计，什么是好的空间？无论是小朋友来画画，还是中学生来现场演奏，这些活动都能让这片空间在他们心目中留下深刻的印象，也通过这些活动让大家认识到把空间做好有怎样的意义。这片区域经历了岁月的冲刷，不免有一些河道、水网、生态被破坏，现在我们要通过各种手段、技术做存量更新，提升环境，这应该成为一种共识。"回到这次长宁区案例展本身，她认为案例展的作品给空间品质提升赋予了很多灵感，绿化空间里不止需要种花草，还可以有灯光，有香味，有城市家具，有很多的可能性。

（本文根据对策展人、相关单位负责人的采访内容整理而成）

Conclusion

"What is the point of SUSAS?" According to Su Liqiong, deputy director of Urban Planning and Natural Resources Bureau of Changning District: "the aim of SUSAS is offering some excellent works for exhibition and public events so that the citizens may better understand what kinds of planning and designs are admirable, and what kinds of spaces are good. The events where, for instance, young kids draw pictures and teenagers perform instruments, may deeply impress the participants and inform them of the significance of space improvement. The rivers, waterways and ecologies in this district have inevitably been damaged through the years, and now we have to renovate the 'stock' and improve our environment with all kinds of methods and technologies – this should be a consensus." And about the Changning site project itself, Su believes that the exhibited artworks have plentifully inspired our efforts to improve urban spaces; and beyond flowers and grasses, "greening" involves many more possibilities: lighting, fragrance, and among others, street furniture.

(This article is edited from interviews with the curators and managers of Changning site project)

连接滨水，邂逅美好——浦东新区实践案例展

Connecting the Waterfront and Encountering Joy — Site Project at Pudong New Area

胡颖蓓
上海城市公共空间设计促进中心 注册城乡规划师

Hu Yingbei
Registered Urban-rural Planner of Shanghai Design
and Promotion Centre for Urban Public Space

主办单位：
上海市浦东新区人民政府
承办单位：
浦东新区规划和自然资源局
协办单位：
上海市城市规划设计研究院、浦东新区妇
女联合会、共青团上海市浦东新区团委员
会、艺仓美术馆
策展人：
奚文沁、卞硕尉、杨帆
展览时间：
2019 年 11 月 8 日—2019 年 11 月 17 日
展览地点：
浦东新区滨江 22 个望江驿

Host:
People's Government of Pudong New Area, Shanghai
Organizer:
Pudong New Area Planning and Natural Resources Bureau
Co-organizers:
Shanghai Urban Planning and Design Research Institute, Pudong
New Area Women's Confederation, Pudong New Area Committee
of the Communist Youth League of China, Modern Art Museum
Shanghai
Curators:
Xi Wenqin, Bian Shuowei, Yang Fan
Time:
8 November 2019 – 17 November 2019
Venue:
22 river view service stations in Pudong New Area

背景：从滨江贯通到品质提升

黄浦江两岸公共空间贯通开放工程 45 公里
涉及全市五个区。其中位于黄江东侧的浦东
新区沿线段（俗称"东岸"）北起杨浦大桥，
南至徐浦大桥，沿途包括新民洋滨江段、上
海船厂滨江段、小陆家嘴滨江段、老白渡滨
江段、外环外三林段，总长 22 公里，是全
市五区中距离最长、工程量最大的。

东岸公共空间贯通开放工作于 2015 年年底
启动；2016 年 2 月，在"众创众规"设计
模式下，来自荷兰、法国、澳大利亚、美国
的五家国际知名设计机构以及社会各界人
士，共同参与了东岸公共空间贯通概念方案
国际征集、青年设计师竞赛、规划调整优化
等工作；2017 年年初，项目开工；至 2017
年底，东岸 22 公里沿线滨江公共空间（以
下简称"东岸滨江公共空间"）实现了规划
设计蓝图中最基本的贯通和开放。2018 年
至 2019 年期间，东岸滨江公共空间进入品

Background: From Waterfront Opening to Space Improvement

The Opening of the 45 km Waterfront Public Space along
the Huangpu River involves five districts of Shanghai, and its
Pudong New Area section on the eastern bank of Huangpu River
(commonly known as "East Bund") which stretches southward
from Yangpu Bridge, through the subsections of Xinminyang,
Shanghai Shipyard, Minor Lujiazui, Laobai Ferry and Sanlin-
beyond-Outer-Ring until Xupu Bridge, is the longest (22km)
one among all those of the five districts and demands the most
efforts.

The Opening of East Bund Public Space began in the end of
2015. In February 2016, in the design mode of "Participatory
Creation and Planning", five renowned international design
agencies from Netherlands, France, Australia and the United
States, and the general public, participated in the events such
as Open Call for Conceptual Design of East Bund Public Space,
Young Designer Competition and planning modification and
optimization. The project entered the construction stage in early
2017, and the basic elements of the planned 22-km waterfront
public spaces on the eastern bank of Huangpu River (hence
"East Bund Public Space") were completed and opened by
the end of the same year. The years of 2018 and 2019 were

实体空间展示分布示意图（图片来源：策展方案）
Map of Exhibition Venues (source: curatorial program)

实践案例展宣传画册示意图（图片来源：策展人）
Site project brochure (source: curators)

质提升阶段，进一步完善公共空间的内涵，增设文化休闲场馆、市民服务驿站等各类公共服务设施；进一步提升滨江可达性，促进公共空间沿纵向通道渗透腹地社区。未来，东岸滨江公共空间将注入更多的产业功能和文化内涵，与浦西四区交相辉映，打造体现上海城市精神的核心场所和世界级滨水区。

呈现：从空间塑造到内涵挖掘

本次浦东新区实践案例展以"连接滨水，邂逅美好"为主题，以近几年浦东新区滨江地区优秀的城市公共空间实践成果为基础，通过一系列由社会各界多方共同举办并组织市民共同参与的公众活动，探讨滨水地区与城市腹地的"连接"，社区与城市的"连接"，探索滨水地区丰富多彩的可能性，展现浦东新区近年来为了让人民生活更美好的不懈努

devoted to improving on the East Bund Public Space, aiming to refine their significance, add culture and leisure facilities, civil service stations and other public amenities, make East Bund more accessible and let public spaces extending towards the hinterland through passageways vertical to the river. In the future, more industrial functions and cultural connotations will be introduced to the East Bund Public Space, and, in collaboration with the four districts on the other side of Huangpu River, create an urban core and a world-class waterfront that embodies the spirits of Shanghai.

Presentation: From Shaping Spaces to Exploring Meanings

The site project of Pudong New Area, under its theme "Connecting the Waterfront and Encountering Joy", is based on the excellent waterfront urban public space projects in Pudong New Area in the last several years. Through a series of public events which are co-held by multiple agents and joined by citizens, it discusses how to connect waterfront with urban hinterland, neighborhoods and the city, explores the abundant

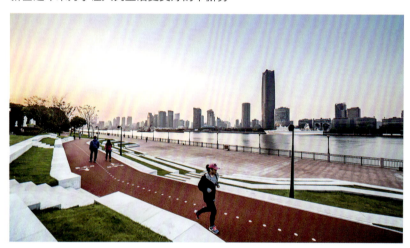

东岸滨江公共空间（摄影：邱岳）
East Bund public space (photograph: Qiu Yue)

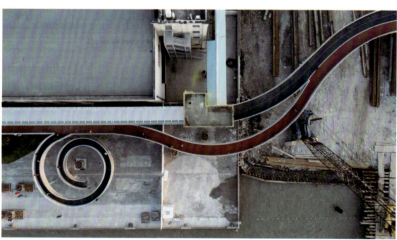

东岸滨江公共空间（摄影：田方方）
East Bund public space (photograph: Tian Fangfang)

2019 年 11 月 12 日东岸滨江漫步（图片来源：活动主办单位）
Stroll the East Bund at 12 November 2019 (source: host)

力，推动东岸滨江公共空间对标世界一流滨水区，不断提升建设和管理水平，吸引更多市民"相遇""相聚"东岸滨江公共空间，向人们传达更加"艺术"的公共空间和更加"美好"的生活方式。

本次实践案例展采用了实体空间展示、主题公众活动两种形式。其中，实体空间展示注重亲身体验，感受滨江公共空间中的建设成果，如东岸滨江公共空间、艺仓美术馆、望江驿、近江社区空间。主题公众活动强调了公共空间在建造和运营过程中的共建共治共享，邀请市民、社会组织、企业、管理部门和专家学者等各方共同参与。

在东岸滨江公共空间游览漫步

作为线形空间，东岸滨江公共空间是本次实践案例展实体空间展示的空间基底。全线通过慢行系统"三道"，即漫步道、跑步道、骑行道，串联起沿江重点区域和重要节点。在轮渡站、河道等断点，通过 12 座充满艺术美感的慢行桥连接。三道时而分开，时而并行，在公共空间中交织，实现了全新的浦江观光方式。人们沿着岸线或漫步，或跑步，抑或骑行，迎着朝阳呼吸新鲜空气，踏访"涅槃重生"的百年老船厂遗址，感受历史、工业与艺术、商业的火花，融入满眼粉色的粉黛乱子草，与沿途怒放的"百草园"合影，饱览浦江两岸的壮丽景色。

本次实践案例展在东岸滨江公共空间的"三道"上，选取从 8 万吨筒仓（2017 上海城市空间艺术季主展场）到东方明珠游船码头的 4 公里漫步道，举办了一场"2019 浦东慢行日——东岸滨江漫步"主题公众活动，邀请沿线陆家嘴街道、塘桥街道、潍坊街道、洋泾街道的近百名市民共同参与，欣赏秋日

and varied possibilities of waterfronts, and demonstrates the recent efforts of Pudong New Area to develop the East Bund Public Space according to the standards of world-class waterfronts, keep improving its constructive and administrative capacities so that the citizens may enjoy ever better lives. The site project is intended to attract more citizens to "encounter" and "meet" at the East Bund Public Space, and convey more "artistic" public spaces and more "enjoyable" lifestyles.

The site project consists of exhibitions and thematic public activities. Exhibitions focuses on intimate experiences through which visitors may personally feel the accomplishments of waterfront public space development, such as East Bund Public Space, Modern Art Museum Shanghai, river view service stations and waterfront neighborhood spaces. Thematic public activities emphasize the participatory elements in the development and operation of public spaces, inviting a wide range of stakeholders such as citizens, social organizations, businesses, authorities and academicians.

Roam the East Bund Public Space

The linear East Bund Public Space is the physical foundation of the exhibitions. The key areas and nodes along the line is connected with three slow-traffic systems – stroll, jogging and cycling– and twelve aesthetical slow-traffic bridges over ferries or rivers. The three lanes, sometimes separated and running in parallel in other times, are interwoven into the public space and constitute a new way to see the sights of Huangpu River. One can wander, run or ride along the river, breathing fresh air and shone by the rising sun, and visits the reinvigorated shipyard of a century's history, feels the sparks of history and industry, art and commerce, delves into the sea of pink muhly grasses, takes a picture with the "gardens of a hundred florals" and be treated with the magnificent sceneries on both banks of Huangpu River.

From all the three lanes of East Bund Public Spaces, the 4km stroll lane between 80,000-ton Silo (main exhibition of SUSAS 2017) and the cruise ferry of Oriental Pearl TV Tower is selected for the thematic public activity "2019 Pudong Day of Slow-traffic – stroll the East Bund". Nearly 100 residents of the adjacent sub-districts of Lujiazui, Tangqiao, Weifang and Yangjing are invited to appreciate the autumn of Huangpu River, experience the accomplishments of Huangpu Opening, and listen to the

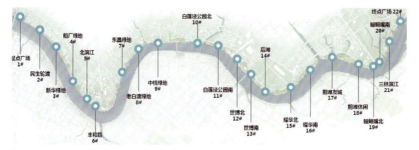

望江驿分布示意图（图片来源：致正工作室）
Map of river view service stations (source: Atelier Z+)

望江驿（图片来源：致正工作室）
River view service stations (source: Atelier Z+)

浦江，体验贯通成果，开发和管理主体的一线工作者、参与沿线设计的建筑景观设计师沿途做了深入浅出的解说。一路上市民你一言我一句，热情高涨，活动结束后仍难掩兴奋之情。

通过专家的现场解读，公众的城市游览不再仅仅停留在"颜值"上，促进其对城市规划理念有了更理性的认知，对城市建设过程中所遇到的问题和困难有了更为深刻的理解和支持；同时，参与活动的市民纷纷表示将邀请他们的家人和朋友前来亲身体验东岸滨江公共空间，还有不少人在活动中结识了有相同兴趣爱好的朋友。

在望江驿停歇充电

在线形的东岸滨江公共空间中，结合区段定位、沿线腹地功能，每 1 公里设置 1 处，统一新建 22 处小尺度社区型公共服务设施点，供滨江活动的人休憩和补给，是不可或缺的落脚点和庇护所，并被冠名为"望江驿"。每处望江驿都是一栋占地面积约 200 平方米的木质结构小屋，由致正建筑工作室主持建筑师张斌设计。小屋背江一侧集合设置了基本便民服务功能，如公共厕所、公共休息室、自动售卖机、雨伞架、信息发布栏等，面江一侧通过大大的落地窗打造开放通透的公共休息厅。天气晴好，温暖透过天窗，屋外阳光明媚，屋内明亮轻快，人们从滨江腹地走来，拾级而上，穿过建筑，面江而坐，江风拂面。

建成开放后，22 座望江驿根据公众的需求升级成为各具主题的文化空间，有安静的书房和热闹的直播室，也有科技感十足的 VR 体验厅和传统邮政文化体验室，还有百姓摄影展、名人名家手稿和珍品展。小小的驿站成为了东岸滨江公共空间中品质提升的标志性景观。它不仅拥有令人赏心悦目的颜值，也有合理的流线和功能安排，给人方便和自在，用温暖的空间承载了人的需求，还有一定的内涵，给人精神上的洗礼。

本次实践案例展将望江驿作为活动服务接待处，发放活动手册和滨江图册，播放宣传视频，提供游客问询。在"东岸滨江漫步"活

lucid interpretations by the front-line staff of developers and administrators as well as the architects and landscapers involved in the waterfront projects. After the frequent discussions and spiritedness during the event, the participating citizens still find hard to hide their passion.

Thanks to the on-site interpretations of professionals, the citizens are allowed to appreciate, beyond the superficial prettiness, the notions of urban planning from a more rational perspective, and be more understanding and supportive towards the difficulties and problems of urban construction. And the participants have shown willingness to invite their family members and friends to personally experience the East Bund Public Space, and not a few of them have become friends with shared interests.

Recharge Yourself at a River View Service Station

22 river view service stations are installed, one for each kilometer of the linear East Bund Public Space and arranged according to the positioning of its section and the functions of its hinterland, where people can rest and refill themselves along the river. These new small-scale community public service facilities, constructed according to an integrated plan, are indispensable stopovers and shelters; and their name in Chinese (Wang Jiang Yi) means literally "Post for Watching Huangpu River". Each service station is a wooden cottage covering an area of about $200m^2$, designed by Zhangbin, the chief architect of Atelier Z+. Basic amenities such as public toilet, lounge, vendor machine, umbrella stand and bulletin board are all placed at the back of each service station, while the open and transparent hall directly faces Huangpu River through French windows. Amidst the sunny warmth and brightness shone through the sunroof, people may come from the hinterland to a service station, follow its steps, walk through the structure, sit down before the river and feel the breezes caressing their faces.

After they are completed, the 22 service stations have been upgraded to distinctively themed cultural facilities. There is a tranquil study, a bustling streaming studio, a high-tech VR shuttle, an experiential old-style post office, a display hall for photographs taken by common citizens, and an exhibition room of manuscripts and rare items of renowned figures. These small service stations have become an iconic landmark for the improvement project of the East Bund Public Space. Besides delightfully appealing appearances, they are convenient and comfortable with their sound arrangement of flows and functions, satisfy the true demands of people in a warm atmosphere, and inspire visitors spiritually with its meanings.

The service stations are designated as Hubs of the site project where event brochures and waterfront maps are distributed, promotional videos are played and advices are given to visitors in demand. In the event of Roam the East Bund, the designer Zhang Bin leads the participants to experience the river view service stations which he has designed, while the accompanying

动中,设计师张斌带领大家亲身体验望江驿,随行的管理和运营主体单位倾听并收集了公众意见,为不断提升服务水平做足功课。

在艺仓美术馆休闲观展

东岸滨江公共空间在更新过程中,保留了一批极具历史价值的工业遗产,如民生码头8万吨筒仓、上海船厂、老白渡煤仓等。这些工业遗产被活化再利用,通风管成了空调管道或电缆收纳器,废弃的钢板被制作成指示牌,粗糙的混凝土支柱成了空间隔断,同时空间被赋予了新的功能,时尚秀、艺术展等文化活动纷至沓来,所有第一次走进这里的人都忍不住惊叹。

艺仓美术馆(图片来源:策展方案)
Modern Art Museum Shanghai (source: curatorial program)

其中,位于张家浜北侧原上海煤运码头区的老白渡煤仓具有一定典型性。2015年是它华丽转身的元年。它在原始状态的基础上通过简单搭建,作为浦东新区实践案例展参与2015上海城市空间艺术季。策展人冯路和柳亦春以"重新装载"为主题,将细腻的舞蹈、影像和声音等当代艺术融入粗砺的工业遗存建筑中,让人们身临其境地意识到历史的价值,探讨工业遗存建筑活化利用变身公共文化空间的意义。自此,原来定位为画廊的艺仓升级为美术馆,正式命名为"艺仓美术馆"。美术馆建筑改造由大舍建筑设计事务所主持建筑师柳亦春负责。设计保留了建筑本体及其北侧绵延数百米的高架运煤长廊,整体嵌入老白渡绿地景观中;建筑外立面以横向线条呼应滨江水面环境,突出原有结构和材质的特征,建筑面积9000余平方米。美术馆由具有一定国际经验的团队运营,在2016年年底开幕后的近两年中举办了近30场展览,其中不乏"神美·米开朗基罗大展""土

representatives of the managing and operating entities listen to and collect the feedbacks of the public, preparing themselves for serving the public ever better.

Relaxed Visit to Modern Art Museum Shanghai

In the renovation of East Bund Public Space, a number of valuable industrial heritages are preserved, such as the 80,000-ton silo of Minsheng Port, Shanghai Shipyard and Laobaidu Coal Bunker, and repurposed: ventilation tubes are turned into AC pipelines or cable management systems, signboards are made out of abandoned steel plates, rough cement columns become room separations. And new functions are assigned to these spaces. Every first-time visitor to the renovated sites must be marveled at the succession of fashion shows and art exhibitions playing out there.

Laobaidu Coal Bunker to the north of Zhangjiabang which used to be part of the coal dock of Shanghai, is a typical example. 2015 is its Year Zero of transformation. With simple add-ons to the original site, the bunker took part in SUSAS 2015. Given the theme "Reload", Feng Lu and Liu Yichun, the curators integrated contemporary art – delicate dances, images and sounds – into the unpolished industrial heritage, leading to a personal realization of the values of history and a discussion about the meaningfulness of transforming industrial heritages into public cultural spaces. Since then, the Art Depot, originally designated as a small gallery, has been upgraded to Modern Art Museum Shanghai. The renovation project of the museum is directed by Liu Yichun, chief architect of Atelier Deshaus. The main structure of the bunker and its overhead coal conveyor that stretches for a few hundred meters on its north are kept intact and embedded as a whole into the landscape of Laobaidu; the facade echoes the river surface with horizontal lines, highlighting what's characteristics about the original structure and materials. The building area of the museum is over 9,000m². In the two years since its opening at the end of 2016, the art museum operated by a team experienced in international projects hosted nearly thirty exhibitions, including international cultural events such as The Divine Michelangelo, Dobuku Civil Engineering and Giorgio de Chirico & Giorgio Morandi: The Shine of Italian Modernism. It is becoming a new architectural, art and cultural landmark of Shanghai. The ground floor and accessory gallery of its main building, as an essential element of the East Bund Public Space, is open to the public and provides all kinds of services, making it another good leisure destination to the south of the Lujiazui Activity Zone.

The "Waterfront Forum", as part of the site project, is held at the museum. Representatives of Shanghai Urban Planning and Natural Resources Bureau, Urban Planning and Natural Resources Bureau of Pudong New Area, Shanghai Urban Planning and Design Research Institute, Shanghai Design and Promotion Centre for Public Space, Lujiazui Administration

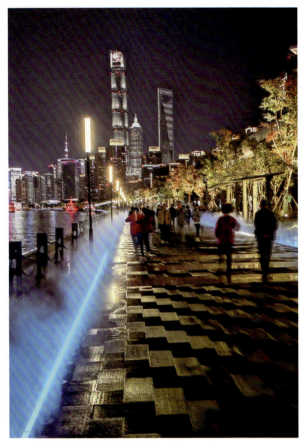

艺仓美术馆长廊改造前后（图片来源：大舍建筑设计事务所，鲍伶俐）
Gallery of Modern Art Museum Shanghai, before and after renovation (source: Bao Lingli, Atelier Deshaus)

2019 年 11 月 29 日"对话滨江"论坛（图片来源：活动主办单位）
Waterfront Forum at 29 November 2019 (source: host)

洋泾生态艺术公园（图片来源：活动主办单位）
Yangjing Eco-art Park (source: host)

潍坊园（图片来源：活动主办单位）
Weifang Garden (source: host)

2019 年 11 月 8 日儿童友好缤纷社区建设推进会（图片来源：活动主办单位）
Promotional Meeting for Building Children-friendly Diversified Community at 8 November 2019 (source: host)

木展"、"基里科＆莫兰迪意大利现代艺术"等国际化世界级艺术活动，并已逐渐成为上海建筑、艺术、文化领域的新地标。建筑主体的底层和附属长廊作为东岸滨江公共空间的重要组成部分，向公众开放，提供了各类服务，也已成为继陆家嘴滨江活动区往南又一处休闲活动的好去处。

本次实践案例展在此举办了一场"对话滨江"论坛。来自市规划资源局、区规划资源局、市城市规划院、市公共空间设计促进中心、陆家嘴管理局、团区委、区文联、区艺术指导中心、区妇儿工委办、区规划管理中心、东岸集团、区各街镇的代表，及艺仓美术馆的建筑设计师和运营管理方等共同参与讨论，聚焦浦东城市公共空间建设和文化发展话题，分享了美术馆在规划、设计、建造等过程中所跨越的重重难关，在工业遗存保留方面的尝试、突破及创新；探讨了东岸滨江公共空间对延续城市文脉、推动城市文化发展的重要意义。

在近江社区空间生活玩耍

东岸滨江公共空间是城市级的"会客厅"。然而贯通后如何将公共、开放的"城市客厅"连接到滨水腹地中的"家门口"，满足社区居民日常生活需求，是更为细致而繁琐的工作。浦东新区缤纷社区计划通过社区口袋公园、社区街巷等一系列项目实现了这一目标。

Bureau, Party Committee of Pudong New Area, Pudong New Area Federation of Literary and Art Circles, Pudong New Area Art Guidance Centre, Working Committee on Children and Women of Pudong New Area, Planning and Administration Centre of Pudong New Area, East Bund Group, sub-districts and towns of Pudong New Area, and architects and administrators of Modern Art Museum Shanghai come together to, focusing on the issues about constructing public spaces and cultural development of Pudong, share their opinions about the difficulties overcome in the planning, design and construction of Modern Art Museum Shanghai, explain its attempts, breakthroughs and innovations in terms of preserving industrial heritages, and discuss the important role that the East Bund Public Space plays in inheriting the cultural veins and promoting the cultural development of Shanghai.

Live and Play in Waterfront Neighborhoods

East Bund Public Space is a city-scale hall. But more minute and complicated tasks await when the opening project is completed, that is, how to connect the public, opened "urban hall" to the "doorsteps" of hinterland and satisfy the daily needs of neighborhood residents. This goal is achieved by the program of Pudong New Area Diversified Community through community pocket parks, community streets and other projects.

In the site project, six typical diversified community projects – Yangjing Eco-art Park, Miaopu Road-Dingshui Road Pocket Park, Dynamic Fushan Road, Rushan Road Vital 102, Weifang Garden, E'shan Road Pocket Park – are presented in forms of physical space and promotional videos played in River View Service Stations. What's more, as the site project focuses on an essential player, children, in the building of communities, a promotional meeting for building children-friendly diversified community, the "Build Diversified Community with Children"

鳗鲡嘴儿童活动空间 [图片来源: TER (岱禾) 景观与城市规划设计事务所]
Manlizui Children Playground (source: Agence TER)

本次实践案例展通过现场的实体空间和望江驿的宣传视频两种形式予以展示，介绍了洋泾生态艺术公园、苗圃路定水路口袋公园、福山路活力街、乳山路活力 102、潍坊园、峨山路口袋公园 6 个具有代表性的项目。此外，活动重点关注了社区营造中的主角——儿童，结合浦东新区儿童友好社区创建，由区规划资源局、区妇儿工委办主办了"缤纷社区你我'童'行"——儿童友好缤纷社区建设推进会，邀请来自花木街道海桐小学的小朋友大胆畅想未来社区，宣读《浦东新区儿童友好社区建设倡议书》。

感悟：从用眼睛看，到用脚丈量，再到用心体会

滨水公共空间连接了"人"

东岸滨江公共空间建设关注"人"的需求，并将"人"和"人"相连。本次浦东新区实践案例展仅仅是东岸滨江公共空间建设运营过程中无数次公众参与活动和工作的一个片段。在建设中，东岸积极向市民、游客问需问计，把一个个不可能变成可能和现实，从系统的角度出发落实细节，真正做到还江于

event is hosted by Urban Planning and Natural Resources Bureau and Working Committee for Women and Children of Pudong New Area, inviting students of Haitong Primary School at Huamu Sub-district to imagine the future of community and announce *Pudong New Area Proposal for Building Children-friendly Community.*

Enlightenment: View by Eye, Measure by Foot and Feel by Heart

Waterfront Public Spaces Connect "People"

The construction of East Bund Public Space focuses on the needs of "people" and connects "people". The Pudong New Area site project is only a segment among the numerous activities and tasks with elements of public participation as the East Bund Public Space is constructed and managed. In the construction phase, inquiries are actively pursued of citizens and visitors about their demands and ideas, thus turning the impossible into the possible and realizations. Details are implemented from systematic perspectives so that the river, its banks and landscapes can be truly returned to the people. Since the opening project is completed, the East Bund Public Space has been continuously responding to people's new demands for life quality, adding various amenities and facilities in service of the people, and providing places where people can meet.

Waterfront Public Spaces Need to Offer More Experience of "Good Life"

After the site project of Laobaidu Coal Bunker in 2015 and the

民、还岸于民、还景于民。在贯通后，持续回应人们对品质生活的新需求，延展丰富的配套功能和贴心的便民服务，为市民提供相互交往的场所。

滨水公共空间需要提供更多"美好生活"体验

从 2015 上海城市空间艺术季的实践案例展老白渡煤仓，到 2017 上海城市空间艺术季的全市主展民生码头 8 万吨筒仓，再到 2019 上海城市空间艺术季的实践案例展东岸滨江公共空间及其周边近江社区空间，东岸滨江公共空间的城市更新实践连续三届参与上海城市空间艺术季。

在历时六年的持续酝酿过程中，人们对于东岸滨江公共空间的关注逐步从点到线，由表及里，并进一步自建设至运营。东岸滨江公共空间要成为让人流连忘返的世界级"城市会客厅"，不仅要"腾"出一片供人观赏的滨水景观和公共空间，更重要的是还要提供丰富多彩的体验，使之成为可以身临其境、乐在其中的交往场所。未来东岸滨江公共空间的运营和管理还应当不断拓宽视野，探索各种途径。

据悉，"十四五"期间，东岸将以关注品质、魅力和人性关怀为发展理念，将进一步推进公共空间品质提升、管理提升以及价值提升，挖掘更多尚未利用的公共空间的潜力。

main exhibition at Minsheng Port 80,000-ton Silo in 2017, this is the third time that the urban regeneration practices of East Bund Public Space presents itself to SUSAS, now with its site project of 2019 which involves, in addition to the East Bund Public Space itself, its adjacent near-river community spaces.

In this continuous development of six years, the focus on East Bund Public Space has gradually grown from dots to lines, from waterfront to hinterland, and from construction to operation. To make itself a world-class "urban hall" to which people would love to come again and again, it is much more important for East Bund Public Space, instead of merely "clearing out" waterfront landscapes and public spaces for people to watch in admiration, to provide a diversified portfolio of experiences for people to be present, enjoy themselves and meet each other. In the future, East Bund Public Space still needs to keep broadening its visions and trying out various approaches in operation and management.

It is reported that in the period of the 14th Five-year Plan, East Bund will adopt the developmental concepts of focusing on quality, charms and humane care, further improving the quality, management and values of public spaces, and utilize more potentials of underused public spaces.

艺术点亮水岸生活

Art Enlightens Waterfront Life

静安区彭越浦河岸景观改造实践案例展

Site Project in Jing'an District: Landscape Renovation of Pengyuepu Waterfront

金江波
上海大学上海美术学院副院长
张承龙
上海大学

Jin Jiangbo
Deputy Director of Shanghai Academy of Fine Arts, Shanghai University
Zhang Chenglong
Shanghai University

主办单位：
上海市静安区人民政府
承办单位：
静安区规划和自然资源局、静安区彭浦镇人民政府
协办单位：
静安区绿化管理中心、静安区河道水政管理所、上海公共艺术协同创新中心
策展人：
金江波、张承龙
展览时间：
2019 年 11 月 2 日—2019 年 12 月 27 日
展览地点：
彭越浦路（汶水路—广中西路）滨河空间

Host:
People's Government of Jing'an District, Shanghai
Organizers:
Jing'an District Planning and Natural Resources Bureau, Pengpu Town Government of Jing'an District
Co-organizer:
Afforestation Management Centre of Jing'an District, River and Water Administration of Jing'an District, Shanghai Public Art Cooperation Centre
Curators:
Jin Jiangbo, Zhang Chenglong
Time:
2 November 2019 – 27 December 2019
Venue:
Waterfront of Pengyuepu Road (Wenshui Road to West Guangzhong Road)

从滨河空间的规划改造到"流域·邂逅"主题

"流域·邂逅——静安彭越浦·社区重塑滨水公共艺术现场"是 2019 上海城市空间艺术季静安区实践案例展。主题"流域·邂逅"是对 2019 上海城市空间艺术季主题的延展和诠释。主展立足于上海市"一江一河"公共空间开发战略，以"相遇"为题聚焦"滨水空间为人类带来美好生活"，"流域"是对"滨水空间为人类带来美好生活"的区域性想象，是对彭越浦滨河空间整体规划改造以来的一次阶段性展示，"邂逅"是"相遇"的深层次参与，寓指大家将在静安彭越浦河的景观区域里，在公共艺术中不期而遇，通过艺术作品与滨河空间互动互融。

彭越浦河是苏州河的一条重要支流，属上海市水利分片蕴南片，河道为南北走向，南端通苏州河，北端与东茭泾首尾相接，通过东茭泾与蕴藻浜相接，静安区内河道长约 6.2

From Waterfront Planning and Renovation to the Theme "Watershed & Encounter"

"Watershed & Encounter — Jing'an Pengyuepu Community Reshaping Waterfront Public Art Site" is the SUSAS 2019 site project in Jing'an District. Its theme "Watershed & Encounter" is an extension and interpretation of the general theme of SUSAS 2019. While the main exhibition is based on the public space developmental strategy of "Huangpu River and Suzhou Creek", focusing on "How Waterfronts Bring Wonderful Life to People" with its theme of "Encounter", "Watershed" is a localized imagination of "How Waterfronts Bring Wonderful Life to People" as an interim demonstration of the comprehensive planning and reshaping of the Pengyupu waterfront, and "Encounter" points to the deep participatory element of meeting to the effect that people can unexpectedly run into public arts among the landscape of Jing'an Pengyuepu, interacting and integrating with the space through art.

Pengyuepu River is a major tributary of Suzhou Creek and part of the hydraulic region of South Wenzaobang River. After diverging from Suzhou Creek, it runs northwards into Dongjiaojing River through which it is connected to Wenzaobang River. Its Jing'an section is about 6.2km in length. The water quality and bank

流域·邂逅—静安彭越浦·社区重塑滨水公共艺术现场
WATERSHED · ENCOUNTER JING'AN PENGYUEPU COMMUNITY RESHAPING WATERFRONT PUBLIC ART SITE

地址信息
Address
彭越浦滨水空间
（汶水路-广中西路）
Riverside Space of Pengyuepu
(Wenshui Road to Guangzhongxi Road)

公共交通
Public Transit
公交79、845、959可达
地铁7号线行知路站（2号口）
Metro Line 7 Xingzhi Road Station (Exit No. 2),
Bus 79, 845, 959 can be reached,
地铁1号线上海马戏城站（4号口）抵达可达
Metro Line 1 Shanghai Circus Station (Exit No. 4)
Can choose

指导单位
Instruction
静安区人民政府
Jing'an District

主办单位
Host
静安区文化和旅游局
Jing'an District Bureau
of Culture and Tourism

承办单位
Organizer
静安区彭浦新村街道办事处
Jing'an District Pengpu Xincun
Sub-district Office
静安区规划和自然资源管理局
Jing'an District Planning and
Natural Resources
Management Office

策展人·团队
Curator
上海大学
Shanghai University
策展人团队
Curatorial Team
章莉莉
Zhang Lili

协办单位
Co-organizers
上海大学美术学院
上海美术学院
Shanghai Academy
of Fine Arts, Shanghai
University

联合工作室
Confidential Team

2019
11.02-12.27

社区重塑·滨水
公共艺术项目计划
Community
Reshaping Public
Art Action Plan

绘本图书出版计划
Picture Book
Illustration Publishing
Exhibition Plan

影像展·黄浦河美
画面影像计划
To Beautify Blocks:
The Micro-Renewal
Exhibition Plan of
Pengyuepu River

社区美育·审美
共享艺术行动计划
Aesthetic Education into
the Community—Master's
Aesthetic Education
Sharing Plan

1.主视觉墙（广中路-花石段）
主体景观艺术（汶石路-汶水段）
综合景观墙
2.彭越浦建设（汶石路-汶水段）
综合景观

——七大美育主题
——儿童美育共享计划
Relax and Play Art—
Children's Aesthetic Education
Participation Project

"流域·邂逅——静安彭越浦·社区重塑滨水公共艺术现场"主视觉海报
Key visual of "Watershed & Encounter — Jing'an Pengyuepu Community Reshaping Waterfront Public Art Site"

公里。随着上海市开展中小河道整治工作以来，彭越浦水质和两岸环境发生巨大的变化，彭越浦（灵石路—汶水路）段两岸原为白遗桥村唐家桥生产队，区域城市化后有 31 家租赁企业在地块内生产经营。2017 年，静安区根据上海市整体规划要求编制了彭越浦河道整治的工程性方案，结合"五违四必"生态环境治理工作要求，对彭越浦（汶水路—灵石路）两岸环境进行大规模集中治理，对十米退界内的违法建筑进行了全面拆除，全部清退租赁企业，拆除违章建筑 51,620 平方米，确保了河道的全面贯通；通过布设生

environment has improved significantly since the initiation of middle-to-small river regulation project in Shanghai. The banks of Pengpuyue River between Lingshi Road and Wenshui Road had been part of Tangjiaqiao Production Team, Baiyiqiao Village, and were leased to 31 industrial and commercial enterprises after the region was urbanized. In 2017, Jing'an District documented an engineering program of Pengyuepu Regulation according to the master plan of Shanghai, and carried out a large-scale and concentrated regulation of the banks of Pengyuepu given the guidelines of ecological environmental improvement, that is, "Five Illegal Actions and Four Necessary Measures". All illegal constructions within 10 meters along the river were demolished – amounting to 51,620m² in building area – and all leasing companies were expelled to make sure that the river course be

彭越浦河改造后的现况（图片来源：策展团队）
Pengyuepu Waterfront after Renovation (source: curatorial team)

态廊道、泵站改造等工程项目消减沿河泵站对河道放江的污染影响；通过改建防汛墙、改造沿岸绿化、布设曝气装置、铺设生态浮床等治理手段进一步改善河道水质和环境面貌，并将沿河十米清理范围交由静安区绿化中心实施景观滨水绿带。

2018 年夏，滨河景观样板绿带建成开放，通过游径、绿化、便民设施等要素建设，打造了河道两岸连续贯通的城市亲水漫步空间和城市活力节点，具有示范意义。滨河空间开放后，周边社区居民越来越多地参与到这片公共区域中来，形成了以生活、休闲、运动为一体的功能性元素，作为近年来静安区具有代表性的滨河空间改造和规划治理成果，我们的策展目的地便选址于此。

从艺术介入到社区美育滋养

公共艺术作用于城市的增量发展是近年来城市化转型发展的急速需求与必然现象。如何形成有成效、有质量的城市品质公共空间，公共艺术无疑就成为一种有效的方法和手段，而"地方重塑"就是公共艺术的核心要义 [1]。基于它的实践内涵，公共艺术则呈现出更加接近并参与公众生活的价值。从这个意义上来看，"地方重塑"就是公共艺术的基本方法，放到我们的策展里，彰显的就是"社区重塑"的意义。

彭越浦河景观区域正是坐落在彭浦镇的基层社区里。社区作为当代中国社会最基本的单

彭越浦滨河改造现状及展场区域设定（绘图：章畅）
汶水路—永和路段东岸：参与式艺术、现场表演区域
汶水路—永和路段西岸：公共艺术·美术馆展示计划
灵石路—广中西路段东岸：彭浦美丽街区围墙彩绘展示
汶水路—灵石路段全段：沿线景观装置活动展示区域
Map of Pengyuepu Waterfront Renovation and Site Project (drawing: Zhang Chang)
East bank of Wenshui Road – Yonghe Road: participatory art and live performance
West bank of Wenshui Road – Yonghe Road: public art and museum plan
East bank of Lingshi Road – West Guangzhong Road: wall painting of Beautiful Pengpu Streets
Both banks of Wenshui Road – Lingshi Road: waterfront landscape installments

1 金江波，潘力. 地方重塑 [M]. 上海：上海大学出版社，2016.

绘本《梦》/ 艺术家：诸葛钧 （图片来源：策展团队）
Dream, illustrated book by Zhuge Jun (source: curatorial team)

公共艺术作品《留鱼》/ 艺术家：刘雪樟 （图片来源：策展团队）
Remaining Fish, public art by Liu Xuezhang (source: curatorial team)

公共艺术作品《涟漪》/ 艺术家：倪垠佳 （图片来源：策展团队）
Ripple, public art by Ni Yinjia (source: curatorial team)

元，是城镇的细胞，也是社会公众美育的现实起点。尝试完善社会功能，关心社区居民精神生活和对文化艺术的需要，凝聚和温暖人心，构建价值认同和再生，推动美育进社区，形成有温度的活态公共空间，进而提升城市人文生活品质，是我们策展的目的。综合以上愿景，我们的策展方向旨在提升彭越浦社区群众的美育和公众参与度，通过社区重塑的方法，最大限度地实现公共艺术的社会性意义。

展线划分与五大板块交叉实现

在展线的划分上，我们把整个展场确定在汶水路至灵石路全段，通过具体的展项把区域空间、创作空间和展示空间充分融合在一起，形成有机协调的展示路径，以公共艺术落地、在地绘本创作、影像艺术表演以及社区居民参与式艺术实践为实现方式，进一步把展项具体划分为五个板块，分别是：公共与在地；驻地艺术工坊及展示；彭浦，社区微更新；社区重塑与居民参与；艺术，无所不在。

在公共与在地板块中，我们所强调的不仅仅

fully opened. Ecological corridor, pump station renovation and other projects alleviated the pollutions by waterfront pumps. Other regulative measures such are reconstructing flood prevention walls, transforming waterfront vegetation, installing aeration equipment and planting artificial floating islands were implemented to further improve water quality and riverside environment. What's more, the strips within 10 meters along the river has been developed into a waterfront landscape belt by Afforestation Management Centre of Jing'an District.

The model waterfront green belt, completed in the summer of 2018, creates an exemplar urban wanderer-friendly waterfront and a vital node of city along continuously open riverbanks with its walking lanes, vegetation and amenities. This waterfront area, since its completion, has increasingly attracted residents of communities nearby to participate in. With its functions of life, leisure and exercise, it is a representative of recent waterfront transforming and regulating efforts of Jing'an, and therefore selected as the site of our project.

From Art Intervention to Community Aesthetic Education

The increasing public arts in cities constitute an urgent demand and a necessary phenomenon as the urbanization progresses rapidly recently. Public art is doubtlessly an effective approach

---

1   Jin Jiangbo and Pan Li, *Local Remodeling* (Press of Shanghai University, 2016).

公共艺术作品《域见·蒙德里安》/艺术家：马俊（图片来源：策展团队）
Mondrian in Field, public art by Ma Jun (source: curatorial team)

是将艺术品放置在社区公共空间，更是将艺术（美育）的种子植根在彭越浦（灵石路—汶水路）滨河空间里，重视一同参与的过程，要由社区居民与艺术家共同定义作品，因而设置了"彭越浦·故事绘本插画创作出版展示计划"，绘本和插画有着叙述内在思维的传播性本质。三位绘本艺术家诸葛钧、李萌、九更先后来到现场，通过他们的体验与感悟，分别创作了绘本《梦》《一》和《世界上唯一的你》。绘本向生活在此地的人们展示和述说着生命百态、世间万象。

驻地艺术工坊及展示是我们团队在介入城市现场和公共艺术创作时长期所坚持和关注的，植根于现场的发声和创作是直面问题的最好方法。通过引入上海大学上海美术学院院校资源，开展校际合作，同时邀请青年艺术家根植于彭越浦，以参与式观察来挖掘区域独特的文化与生机，最终创作出十二组公共艺术作品。这些作品分别为三所专业艺术高校（上海大学上海美术学院、天津美术学院、四川美术学院）的年轻艺术家创作。作品紧扣滨水空间中的"社区重塑"，表达形

and method to create effective and high-quality urban public spaces, and the core concept of public art is "local remodeling".[1] Given its practical implications, public art demonstrates values that are accessible to and participating in the life of people. In this sense, "place reshaping" is the basic method of public, and in terms of our exhibition, it demonstrates the significance of "local remodeling".

The Pengyuepu River Landscape Area is located at the grassroots communities of Pengpu Town. Community, as the basic unit of modern Chinese society, is the cell of city and the practical beginning of public aesthetic education. Our site project aims to refine the functions of our society, cater to the citizens' demands for spiritual life, culture and art, consolidate and warm people's hearts, construct the identification and revitalization of values, promote community-based aesthetic education, create warm and dynamic public spaces and hence improve the cultural life of city. Given these visions, our curatorial approach points to aesthetic education and public participation of the members of Pengyuepu communities, and tries to maximize the social significance of public art through community remodeling.

Sections and Five Panels Intercrossed

As the map of site project shows, the venue is distributed among the entire banks between Wenshui Road and Lingshi Road where physical spaces, creative spaces and exhibitive spaces

墙体彩绘《海上浮生记》（图片来源：策展团队）
Lives at Sea, wall painting (source: curatorial team)

儿童美育参与计划现场（图片来源：策展团队）
The Participatory Art Education Program for Kids
(source: curatorial team)

式多元丰富，以公共景观装置、地景绘画、声音交互装置等形式沿河展示。

彭浦，社区微更新是策展团队承担的两条"美丽街区"的设计实施项目，两条路分别是彭越浦路（广中西路—灵石路）和彭越浦路（灵石路—汶水路）。我们的设计定位是沿河街道的综合整治以及总体环境的提升，围绕彭浦镇的核心及周边区域，从城市空间导视与公益宣传、滨水空间艺术营造等方面选取六个子更新项目。其中，对彭越浦路（广中西路—灵石路）围墙改建和彩绘设计，出乎意料地成为了今天静安区的网红打卡地，沿河道503米的主题创作彩绘墙在国内也是少见的。彩绘的墙体是彭浦镇工业厂房遗存的一段，墙体外围破损脱落严重，为了加固和

展览开幕式现场（图片来源：策展团队）
Opening ceremony of the exhibition
(source: curatorial team)

美化墙体，我们设计了一幅题为《海上浮生记》的墙体彩绘，勾勒出生活在海上的居民生活百态，既与彭浦亲水步道景观充分融合，又表达了在地居民与自然和谐共处的美好寓意。全图以"水"为基础元素（浪、游鱼）贯穿全篇，契合海上流域的主题。

are thoroughly mingled to constitute organically coordinated demonstrative approaches in the forms of public art installment, on-site picture book creation, visual show and participatory art performance. The exhibited items are divided into five panels: Publicity and Locality; Local Art Workshop with Demonstration; Community Micro-renewal of Pengpu; Community Remodeling and Citizen Participation; and Art Everywhere.

In the panel of Publicity and Locality, we emphasize that planting seeds of art (aesthetic education) in the waterfront fields of Pengyuepu (Lingshi Road-Wenshui Road) is more important than installing artworks in community public spaces. The idea of public participation and allowing citizens to define artworks together with artists has made for the "Illustrated Story Book Creation, Exhibition and Publishing Program of Pengyuepu". Picture books and illustrations are natural media for expressing inner thoughts. Three illustrators, Zhuge Jun, Li Meng and Jiugeng, have come to the site and, with their experiences and feelings, created their works here: *Dream*, *One* and *You are Unique in the World*. These illustrated books demonstrate and tell the multitude of lives and events to local residents.

Local Art Workshop with Demonstration has been our consistent focus of urban scene intervention and public art creation. Scene-rooted voicing and creating is the best way to confront problems. By introducing the resources of Shanghai Academy of Fine Arts, Shanghai University, collaborations between multiple institutions and inviting young artists to settle in Pengyuepu, 12 sets of public artworks are completed as participatory observations of the unique local culture and vitality. These works are created by young artists from three art institutions – Shanghai Academy of Fine Arts, Shanghai University, Tianjin Academy of Fine Arts and Sichuan Fine Arts Institute. Closely attached to the theme of "community remodeling" in waterfront spaces, they decorate the banks in various forms such as public landscape, land art and among others, voice-based interactive device.

Community Micro-renewal of Pengyuepu consists of two "Beautiful Street" projects designed by the curatorial team, that is, Pengyuepu Road (West Guangzhong Road – Lingshi Road) and Pengyuepu Road (Lingshi Road – Wenshui Road). With the vision of comprehensive regulation and environmental improvement of waterfront streets in the centre and periphery of Pengpu Town, we have selected six regeneration projects about urban signage, public announcement and waterfront art. Now the reconstructed and painted walls in Pengyuepu Road (West Guangzhong Road – Lingshi Road), with its 503m of waterfront thematic wall paintings rare in China, have unexpected become a popular Instagram-worthy spot of Jing'an District. These walls used to belong to an old factory in Pengpu town, hardly intact and heavily eroded, so to sake of safety and beauty, we put on them a coloured painting *Lives at Sea*, depicting the various living conditions of people residing at sea. The painting is fully

在社区重塑与居民参与板块中，透过艺术能量浸润彭越浦社区，共同激荡出具有在地生态意识、人文思维、环境关照的艺术创作是我们的策展初衷。如何让社区居民融入在地环境，感受滨水现场，共同发声，是我们持续关注和探讨的。策展上，我们尝试不断调整创作者和参与者的节奏与姿态，希望从总结中发展出不同形态的社区实践，构建新的社区文化艺术平台。儿童美育参与计划开启了社区孩子的美学教育。通过美育滋养课程，以多媒体互动、DIY 动手创作、游戏体验等方式，启发孩子对艺术的认识与兴趣。而美育进社区——大家美育分享计划的设置，是让艺坛大家、名家美育走进社区，关心社区居民精神生活和对文化艺术的需要。我们想做的不仅仅是将艺术品放置在社区公共空间，而是将艺术的种子"种植"在彭越浦（汶水路—广中西路）滨河空间里，社区居民与艺术家们一同参与，共同定义作品的意义。

艺术无所不在，是通过结合科技、肢体展演、自然环境，呈现公共艺术多元的面貌，带动彭越浦社区居民对美育的认知与想象。新媒体艺术秀《形》是在上海大学上海美术学院数码艺术系和上海戏剧学院数字媒体艺术系联合媒体艺术课程上呈现的。课程深入地探讨了如何将媒体艺术课程转化为艺术展演表现，以及如何将它搬进社区。其通过演出搭建一次美妙的相遇——滨水空间与公众群体的邂逅，年轻一代与年长一代的邂逅，城市老旧空间与公共艺术的邂逅，用新媒体技术拓展了舞台表演的原有形式，结合了AR&VR、数字体感侦测、实体装置体验、游戏及影像动画等多种表现形式，拉近了人与人之间的距离，吸引社区居民从家中走入滨水空间，让大家充分体验新媒体时代下的艺术形式。

一次极有价值的尝试

在"流域·邂逅"的推进过程中，"地方重塑"作为公共艺术的核心价值始终贯穿我们的策展全过程。这个过程需要与众多的社区机构和个人合作。我们希望共同探讨滨水社区中的公共空间、文化艺术等方面的可能性，

incorporated into the waterfront promenade of Pengpu Town, and conveys the admirable idea of local human-nature harmony. Echoing the theme of "Watershed", water is the basic thread running throughout the painting in the forms of fish and waves.

The panel of Community Remodeling and Public Participation aims to infiltrate the energy of arts through the Pengyuepu community, and inspire creativity illuminated by localized ecological awareness, cultural idea and environmental insight. It has consistently been our focus and issue to incorporate people with the local environments they found themselves in, get people to feel waterfront scenes and make their voices together. In terms of curation, we keep calibrating the attitudes and paces of both creators and participants, and hope to construct a new community cultural and art platform based on various community actions as we summary and organize the results. The Participatory Art Education Program for Kids marks a beginning of aesthetic education for the young members of Pengyuepu, arousing their understanding and interest of art through multi-media interaction, DIY and games. The Aesthetic Education in Community: Artists Communication Program invites accomplished artists into Pengyuepu and caters to the spiritual life and needs for art and culture of citizens. What we want to do, in addition to installing artworks in community public spaces, is "planting" seeds of art into the waterfront of Pengyuepu Road (Wenshui Road – West Guangzhong Road) where citizens and artists create and define the meanings of artworks together.

The panel of Art Everywhere encourages the citizens of Pengyuepu to conceive and imagine aesthetic education through the wide range of public art that involves technology, body performance and natural environment. The new media show Form is presented on a united media art course by Department of Digital Art, Shanghai Academy of Fine Arts, Shanghai University and Department of Digital Media and Art, Shanghai Theatre Academy. The course offers a thorough discussion of how to translate a media art course into an actual performance, and move it into a community. The show establishes a marvelous case of encounter – encounter between waterfront spaces and citizens, the younger and elder generations, old urban spaces and public art. New media technologies are employed to expand the forms of performance with AR&VR, motion detection, experiential installment, gaming, video and animation, and bring people closer, attract residents to come out to the waterfront and fully experience the arts in the era of new media.

An Extremely Valuable Experiment

"Local Remodeling", as the core value of public art runs throughout the curation of "Watershed & Encounter", a process in which collaboration of many organizations and individuals of the community is indispensable. We hope to discuss the

媒体艺术展演《形》（图片来源：策展团队）
Form, media art show (source: curatorial team)

媒体艺术展演《形》（图片来源：策展团队）
Form, media art show (source: curatorial team)

尝试着合作创建一个"重塑化社区"，进一步建立公共艺术与社区之间的信任和更持久的共存模式。

完善公共艺术的参与与重塑，需要在一个地方，建立长久而活性的"地方重塑"机制，不仅是对地方文脉或习性的深层改良，也是确保实现"使用者自觉"的有效途径。在公共空间中倡导民主、开放、交流、共享的精神与态度以及互动、参与的理念，更多地思考内在的人文关怀，包括地域生态的优化、生活品质的改善、人的幸福指数提升、和谐的人际关系等诸方面指标，是检验公共艺术多元的评价体系的样本。

从这个意义说，"流域·邂逅——静安彭越浦·社区重塑滨水公共艺术现场"是一次极有价值的尝试。

possibilities of public space, culture and art in waterfront communities, and attempt to build a "remodeled community" together so that a more trusted and sustainable co-living model between public art and community may be further established.

A permanent and energetic "local remodeling" mechanism is necessary for the participation and remodeling of public art in a place. It is not only a deep reform of local culture or habits, but an effective approach to establish "user awareness". A sample evaluation system of how diversified the public arts in a place are, consists of promoting a democratic, open, communicative and sharing attitude and an interactive, participatory mindset, and thinking more about inner humanistic care including improving local ecology, life quality, happiness and wellbeing, and among others, harmonious interpersonal relationship.

In this sense, Watershed & Encounter — Jing'an Pengyuepu Community Reshaping Waterfront Public Art Site is an extremely valuable experiment.

普陀区 M50 创意园实践案例展

Site Project at M50 Creative Park, Putuo District

褚欣
普陀区规划和自然资源管理局

Chu Xin
Urban Planning and Natural Resources Bureau of Putuo District

主办单位：
上海市普陀区人民政府
承办单位：
普陀区规划和自然资源管理局、普陀区长
寿路街道办事处
协办单位：
上海 M50 创意园、天安（中国）投资有限
公司、上海音乐学院数字媒体艺术学院
策展团队：
上海交通大学城市更新保护创新国际研究
中心、上海安墨吉建筑规划设计有限公司
展览时间：
2019 年 11 月 16 日—2019 年 12 月 8 日
展览地点：
M50 创意园，莫干山路 50 号

Host:
People's Government of Putuo District, Shanghai
Organizers:
Putuo District Planning and Natural Resources Bureau,
Changshou Road Sub-district Office
Co-organizers:
M50 Creative Park, Tian An China Investments Company
Limited, Department of Digital Media and Art of Shanghai
Conservatory of Music
Curatorial Teams:
Shanghai Jiaotong University International Research Centre for
Creative Urban Regeneration and Protection, AMJ Shanghai
Time:
16 November – 8 December 2019
Venue:
M50 Creative Park, 50 Moganshan Road

2019 年上海城市空间艺术季普陀区实践案例展于 11 月 16 日至 12 月 8 日在普陀区苏州河畔的 M50 创意园举办。本次活动由普陀区人民政府主办，普陀区规划和自然资源局、普陀区长寿路街道办事处承办，上海 M50 创意园、天安（中国）投资有限公司、上海音乐学院数字媒体艺术学院协办。策展团队为上海交通大学城市更新保护创新国际研究中心和上海安墨吉建筑规划设计有限公司。

案例展主题为"从水岸滨河到活力普陀——文化点亮苏河"，紧紧围绕上海市委市政府紧密部署的"一江一河"公共空间开发战略，展示了普陀区规划与发展苏州河的战略与具体实施成果。本次普陀区实践案例展将主展览选址定于苏州河长寿第一湾的 M50 创意园，结合普陀苏州河段已完成贯通工程的南段（西康路桥—武宁路桥南岸），整体呈现普陀苏河 21 公里岸线的历史与未来。

展览由"从水岸滨河到活力普陀——文化点

The SUSAS 2019 site project of Putuo District is held in M50 Creative Park by Suzhou Creek lasting between 16 November and 8 December. This project is hosted by People's Government of Putuo District, organized by Urban Planning and Natural Resources Bureau of Putuo District and Sub-district Office of Changshou Road, and co-organized by M50 Creative Park, Tian An China Investments Company Limited and Department of Digital Media and Art of Shanghai Conservatory of Music. The curators are Shanghai Jiaotong University International Research Centre for Creative Urban Regeneration and Protection and AMJ Shanghai.

The site project, with its theme of "From Waterfront to Dynamic Putuo – Culture Enlightens Suzhou Creek" and close alignment with the public space developmental strategy "Suzhou Creek and Huangpu River" implemented by the Party Committee and Municipal Government of Shanghai, demonstrates the strategy and accomplishments of Putuo District in planning and developing Suzhou Creek. Its primary venue is selected as M50 Creative Park at Suzhou Creek First Bay, Changshou Road, and together with the completed southern section of Suzhou Creek connection project in Putuo District (Xikang Road Bridge – southern bank of Wuning Road Bridge), offers an integral demonstration of the history and future of the 21km Suzhou

M50 创意园现状与昔日景象
M50 Creative Park: past and present

M50 创意园鸟瞰
Bird view of M50 Creative Park

亮苏河"开幕式论坛、普陀苏河 21 公里岸线空间规划主旨展、"听岸"上音数媒学院城市空间声音景观艺术主题展、最美一公里"共生"——天安·千树城市公共空间艺术主题展等内容组成，同时包括丰富多彩的市民活动，如长寿街道老建筑寻访、M50 艺术季、M50 艺术市集等。

Creek section in Putuo District.

The site project consists of the opening ceremony of "From Waterfront to Dynamic Putuo – Culture Enlightens Suzhou Creek", the Keynote Exhibition of the Waterfront Planning for the 21km Suzhou Creek section in Putuo District, an urban space soundscape exhibition "Listen to Banks" by Department of Digital Media and Art, Shanghai Conservatory of Music, a public space art exhibition "Symbiosis of the Most Beautiful Kilometer — Tian An & A Thousand Trees", and a diversified combination of public activities such as Search for Historical Buildings in Changshou Road, M50 Art Season and M50 Art Fair.

规划主旨展
Keynote planning exhibition

规划主旨展
Keynote planning exhibition

普陀苏河 21 公里岸线景观
Landscape along the 21 km waterfront of Suzhou Creek in Putuo District

"听岸"上音数媒学院城市空间声音景观艺术主题展
The urban space soundscape exhibition "Listen to Banks" by Department
of Digital Media and Art, Shanghai Conservatory of Music

策展主题阐述

展示内容简介

"从水岸滨河到活力普陀——文化点亮苏河"开幕式论坛于 11 月 16 日在 M50 创意园举行，普陀区常委、副区长韩金华，上海市公共空间设计促进中心主任徐妍，上海市规划和自然资源局风貌处副处长许晴，普陀区规划和自然资源局局长彭波，同济大学国家历史文化名城研究中心主任阮仪三教授等出席开幕活动。

学术论坛环节，同济大学教授张松，艺术评论家、上海大学艺术研究院副院长李晓峰，上海交通大学教授王林，M50 文化创意发展有限公司总经理周斌依次作主题演讲。而后，园区管理方、文化遗产保护领域的专家以及艺术家，以苏州河公共空间景观的发展、M50 创意园的保护更新与创新研究成果为基础，共同探讨城市滨水活力、历史风貌与工业遗产、艺术产业与公共空间的关系，回顾苏州河工业历史文化遗产保护与当代艺术在此生长发育的历程，探讨、交流普陀苏河滨水空间城市更新的思想，并展望普陀苏州河的未来。著名文艺评论家吴亮，中国当代艺术国际知名艺术家谷文达，纽约大学教授、画家张健君，余留地建筑事务所创始人岳峰，上海交通大学设计学院副教授张海翱等嘉宾受邀出席。

本次活动的主旨展为"迈向令人向往、有温度的世界级滨水空间——普陀苏河 21 公里岸线空间规划"，紧紧围绕上海市委市政府紧密部署的"一江一河"公共空间开发战略，展示了普陀区规划与发展苏州河的战略与具体实施成果。苏州河普陀段岸线全长 21 公里，占市区段总长 50%，沿岸 19 个断点将在年底全部打通。主旨展围绕苏河八景规划、断点节点景观提升、苏河历史与未来等内容，以纪录片、模型展示、空间体验等方式进行呈现。

"听岸"上音数媒学院城市空间声音景观艺术主题展作为 2019 上海城市空间艺术季普陀区分展区的主题展之一，紧扣"文化点亮苏河"的主题，秉持声音景观的基本内涵，

Theme Interpretation

Contents of Site Project

At 16 November, the opening ceremony and forum of "From Waterfront to Dynamic Putuo – Culture Enlightens Suzhou Creek" is held at M50 Creative Park. It is attended by Han Jinhua, member of Standing Committee of Communist Party of China of Putuo District and deputy mayor of Putuo District; Xu Yan, director of Shanghai Design and Promotion Centre for Urban Public Space; Xu Qing, deputy director of scenery office, Shanghai Urban Planning and Natural Resources Bureau; Peng Bo, director of Urban Planning and Natural Resources Bureau of Putuo District; and among others, Professor Ruan Yisan, director of Research Centre for National Historical and Cultural Famous Cities, Tongji University.

The keynote speakers of the academic forum are Professor Zhang Song of Tongji University; Li Xiaofeng, art commentator and deputy director of Art Research Institute, Shanghai University; Professor Wang Lin of Shanghai Jiaotong University; and Zhou Bin, general manager of M50 Culture and Creativity Development Limited Company. After the keynote speeches, administrators of M50 Creative Park, professionals of cultural heritage protection and artists, based on the landscape development in Suzhou Creek public spaces and the M50 Creative Park's achievements in protective regeneration of heritages and creative study, discuss the topics about the vitality of urban waterfronts, historical sceneries and industrial heritages as well as the relationship between art industry and public space, review the process of how industrial and cultural heritages along Suzhou Creek are protected and how contemporary arts are rooted here, exchange their ideas about urban regeneration in the Putuo section of Suzhou Creek, and envision the future of Suzhou Creek in Putuo. The forum also invites honored guests such as Wu Liang, a famous art and cultural commentator; Gu Wenda, a Chinese contemporary artist of international reputation; Zhang Junjian, painter and professor of New York University; Yue Feng, founder of Design Reserve; and Zhang Hai'ao, associate professor of School of Design, Shanghai Jiaotong University.

The Keynote Exhibition "Towards an Admirable and Warm World-class Waterfront – Waterfront Planning for the 21 km Suzhou Creek section in Putuo District", closely aligned with the public space developmental strategy "Suzhou Creek and Huangpu River" implemented by the Party Committee and Municipal Government of Shanghai, demonstrates the strategy and accomplishments of Putuo District in planning and developing Suzhou Creek. The Putuo section of Suzhou Creek, 21 km in length, constitutes 50% of the river within the urban districts of Shanghai and all of its 19 dead-ends will be connected by the end of 2019. The Keynote Exhibition shows the planning of Eight Sceneries of Suzhou Creek, Landscape Design of Dead-ends and Waterfront Nodes and, among other others, History

最美一公里"共生"——天安·千树城市公共空间艺术主题展
Urban public space art exhibition "Symbiosis of the Most Beautiful Kilometer – Tian An & A Thousand Trees"

M50 创意园地块滨水景观步道
Waterfront Landscape Promenade in M50 Creative Park

以大型装置的形式，将声音景观与城市空间的公共艺术相结合，以苏州河流域节段为装置基本形态，以错综复杂的钢管，象征水流与声波的传递途径，以五个管道输出端口，象征五个不同形态的岸口，而不同形态的岸口，则引导不同的倾听角度，五个作品共同构筑起苏州河沿岸属于上海的声音记忆和听觉景观。

最美一公里"共生"——天安·千树城市公共空间艺术主题展作为 2019 上海城市空间艺术季普陀区分展区的主题展之一，旨在探讨在城市发展进程中，街头涂鸦艺术与公众日常空间形成的共生共融的关系。与此同时，"共生"也体现出莫干山路所处的苏州河畔，作为上海都市文明发展的缩影，在城市更新进程中，始终秉承着历史脉络与现代文明共生交汇的建设方式。在未来，这里也将通过更多具有前瞻性与开放性的艺术介入，与苏州河的百年历史文化展开跨时空对话，呈现先锋气息与情怀记忆共生的状态。

案例阐述及评论

2018 年 10 月起，普陀区启动苏州河岸线整体品质提升和贯通工程。苏州河普陀段岸线长达 21 公里，原苏州河岸线功能单一、利用率不高、可达性不强、配套设施不足，为了提升苏州河岸线整体品质，普陀区进行了沿岸整体景观设计、大力推进断点贯通，在条件成熟的岸线率先进行景观步道的实施，形成了多处标志性苏州河景观示范段，包括M50段、苏堤春晓段、长风商务区段等，已成为社区居民和公众喜闻乐见的高品质公

and Future of Suzhou Creek in the forms of documentary video, physical model and space experience.

The urban space soundscape exhibition "Listen to Banks" by Department of Digital Media and Art, Shanghai Conservatory of Music, as a thematic exhibition of SUSAS 2019 in Putuo District, closely follows the theme "Culture Enlightens Suzhou Creek" and the basic concept of soundscape, incorporates soundscape with urban public art by large installments whose basic forms derive from the sections of Suzhou Creek Watershed. The maze of steel tubes symbolizes the ways in which water and sound waves propagate. The five outputs symbolize the five types of ports which in turn directs different listening perspectives. Put together, the five artworks constitute a sound memory and acoustic landscape of Suzhou Creek waterfront that truly belongs to Shanghai.

The urban public space art exhibition "Symbiosis of the Most Beautiful Kilometer – Tian An & A Thousand Trees", as a thematic exhibition of SUSAS 2019 in Putuo District, aims to discuss the co-living and fusing relationship between street graffiti and daily lives of the public with the development of cities. In the meantime, "symbiosis" reflects that the Moganshan Road section of Suzhou Creek waterfront, as an epitome for the development of the metropolitan Shanghai, is consistently constructed in such a way that its histories and the modern civilization co-exist and promote each other as it undergoes regeneration. In the future, more forward-looking and open art will be introduced to start dialogues with the century-long history and cultures of Suzhou Creek, demonstrating a symbiosis between experimentalism and sentimental memories.

Interpretation and Commentary

In October 2018, Putuo District initiated its comprehensive improvement and opening project of Suzhou Creek waterfronts. The 21 km Suzhou Creek waterfronts in Putuo District used to functionally monotone, underutilized, inaccessible and lacking in amenities. To generally improve its Suzhou Creek waterfront, Putuo District has carried out holistic waterfront landscape designs and energetically opened the dead-ends; and has first constructed landscape promenades where conditions are suitable, resulting in multiple iconic and demonstrative sections of Suzhou Creek like M50, Sudi Chunxiao Neighborhood and Changfeng Ecological Business District, which have become high-quality areas for public activities among locals and the wider citizenry. Thirteen dead-ends are opened by October 2019, including the entire waterfront between Wuning Road and Xichang Road. The remaining dead-ends will be opened with the construction of second-tier flood prevention walls, so that in addition to safety, better landscape of Suzhou Creek shall be introduced for everyone to enjoy. After the dead-ends are opened, the tasks for the Putuo Section of Suzhou Creek in the next year will be further improving landscape, introducing more

共活动区域。截至 2019 年 10 月，苏州河普陀段已打通 13 个断点。其中，武宁路至西昌路沿河段全部开放。剩下的断点，将结合二级防汛墙建设进行打通，在实现防汛安全功能的同时，提升景观品质，让人人都可以看到苏州河。打通断点后，苏州河普陀段将在下一年进一步提升景观品质，注入更多人文元素，规划 6 个游船码头，串联周边主要景点，形成艺术创意集聚区、文化高效聚集区，在苏州河畔打造"八大景观"。

M50 创意园地块，作为苏州河普陀区段重要节点，其滨水空间改造将成为苏河岸线景观承上启下的重要展示界面。目前，M50 创意园已经打破围墙，逐步转型成全开放式街区。

活动感想

"苏州河"对上海工业文明的发展有着极其重要的作用和影响。此次 2019 城市空间艺术季，将"从水岸滨河到活力普陀——文化点亮苏河"作为普陀区案例展主题，对于苏州河普陀区段整体围绕"一江一河"公共空间开发战略做出了集中展现，展示了普陀区规划与发展苏州河的战略与具体实施成果。

在开幕式论坛上，相关专家探讨、交流普陀苏河滨水空间城市更新的思想，并展望普陀苏州河的未来。此外，M50 城市空间艺术季更是以影像作品、规划模型、艺术装置等多种形式，在 M50 这个集聚工业、文化、艺术、交流于一体的创意园区，整体呈现了普陀苏河 21 公里岸线的历史与未来。一个城市的记忆，一代人的乡愁，"一厂一回忆"……上一代人的记忆与乡愁也许就寄托于类似 M50 这样的老厂房，这些工业文明留下的印迹都值得保护、探索和研究。

cultural elements, planning for six cruise docks, connecting the major scenic spots in the nearby, forming clusters of creative and productive cultures and creating "Eight Sceneries" along Suzhou Creek.

M50 Creative Park, as a key node of Suzhou Creek in Putuo District, will present its transformed waterfront spaces to be an important demonstrative interface of Suzhou Creek landscape. It has already been breaking down walls and turning itself into a fully open block.

Feelings and Reflections

Suzhou Creek used to an extremely powerful influence on the development of modern industries in Shanghai. The site project, with its theme of "From Waterfront to Dynamic Putuo – Culture Enlightens Suzhou Creek" and close alignment with the public space developmental strategy "Suzhou Creek and Huangpu River" implemented by the Party Committee and Municipal Government of Shanghai, demonstrates the strategy and accomplishments of Putuo District in planning and developing Suzhou Creek.

At the forum of the opening ceremony, professionals have discussed and shared their ideas about urban regeneration in the Suzhou Creek waterfronts of Putuo District, and envisioned the future of Suzhou Creek in Putuo. What's more, the SUSAS site project at M50, a creative park that brings industry, culture, art and communication together, integrally exhibits the history and future of the 21km Putuo section of Suzhou Creek in the forms of video, image, planning model and installment art. Memories of a city, nostalgia of a generation, "the recollections unique to every factory" ... Maybe these memories and sentiments of the passing generation are embodied in old plants like those of M50, and all the impressions left by our industrial civilization are worthy of protection, exploration and study.

西岸时刻——徐汇区上海西岸实践案例展

Moments of West Bund — Site Project at West Bund, Xuhui District

西岸文化艺术季

Art' West Bund

主办单位：
上海市徐汇区人民政府
承办单位：
徐汇区规划和自然资源局、徐汇区文化和
旅游局、西岸集团
协办单位：
光启文化
策展团队：
西岸文化艺术季
展览时间：
2019 年 11 月
展览地点：
上海西岸

Host:
People's Government of Xuhui District, Shanghai
Organizers:
Xuhui District Planning and Natural Resources Bureau, Xuhui District Culture and Tourism Bureau, Shanghai West Bund Development (Group) Co., Ltd.
Co-organizer:
Guangqi Culture
Curatorial Team:
Art' West Bund
Time:
November 2019
Venue:
West Bund, Shanghai

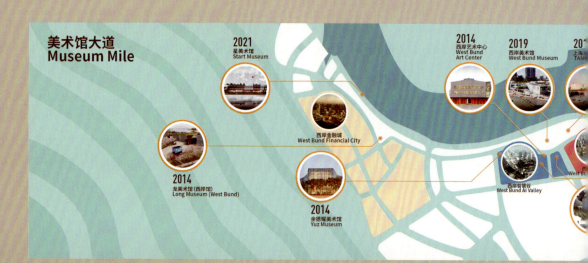

2019 年 11 月 5 日，伫立于徐汇滨江沿岸的西岸美术馆正式开馆。当天下午，在这个由建筑大师大卫·奇普菲尔德设计的建筑里，在各界媒体、文化界人士的见证下，法国总统马克龙与西岸艺术委员会主席、徐汇区区长方世忠共同为西岸美术馆与法国蓬皮杜中心五年展陈合作项目揭幕。这一合作不仅是中国本土美术馆首次与法国顶级美术馆共同探索运营文化机构，也是当代艺术领域级别最高、周期最长的中外文化交流合作项目。

以此为开端，西岸巅峰艺术月拉开帷幕，随后串联西岸沿线 3 个公共艺术作品、7 个建筑景观、9 个工业遗迹地标的 2019 上海城市空间艺术季徐汇区实践案例展，汇聚全球 110 家画廊、800 多位大师力作的西岸艺术与设计博览会、海内外 30 多家知名院校联手打造的艺术与设计创新未来教育博览会等 30 场展览、120 场活动，在区域内的 17 个场馆里轮番登场，西岸演绎了自己的"西岸时刻"。

时间简史——上海西岸艺术实践

滴水穿石，功不断。如果将西岸美术馆的开馆作为一个时间维度的节点，标志着文化对上海城市空间的影响力，将西岸美术馆视为上海与世界平等对话的文化大平台和文化新地标，西岸则以兼容并包的姿态正在成为"世界的会客厅"，西岸在此之前做出的艺术实践无疑是不可忽视的重要累积。

On 5 November 2019, West Bund Museum, standing on the bank of Xuhui Riverfront, was officially open. In the building designed by master architect David Chipperfield, that afternoon witnessed, with media and members of cultural circles present, the inauguration of the five-year Centre Pompidou × Chinese Native Art Museum West Bund Museum Project attended by Emanuel Macron, the French president and Fang Shizhong, the chairman of West Bund Art Committee and the mayor of Xuhui District. This collaborative project is the first attempt of a Chinese art museum to co-operate a cultural institution with a top-tier French gallery, and the most lasting and high-level cultural collaboration program between Chinese and foreign agents in the field of contemporary art.

The inauguration marked the beginning of West Bund Time. The SUSAS 2019 site project at Xuhui which included 3 public artworks, 7 architectural landscapes and 9 industrial heritages, the West Bund Art & Design where the works of 110 galleries around the world and more than 800 artists were exhibited, 30 exhibitions including Art & Design Education Future Lab as the result of efforts by over thirty renowned education institutions in China and abroad, and 120 events successively took their places in one or another of the seventeen institutions, shining the "moments of West Bund".

A Short History of Art Practices in West Bund, Shanghai

Continuous efforts make for great accomplishments. If we consider the opening of West Bund Museum as a milestone along the axis of time that represents cultural influences on urban spaces of Shanghai, if we see the West Bund Museum as a grand cultural platform and new cultural landmark where Shanghai dialogues with the world equally, and that West Bund is becoming an inclusive "Hall of World", then it is doubtless that previous art practices in West Bund have laid a foundation which

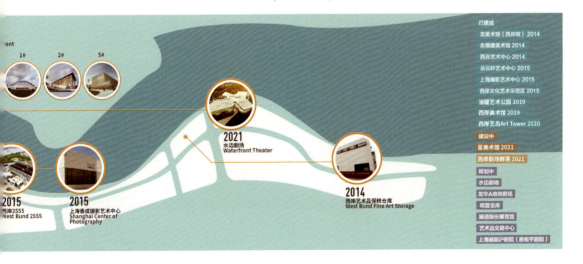

西岸美术馆大道及历史沿革
Historical map of West Bund Museum Avenue

西岸美术馆
West Bund Museum

2012 年西岸音乐节的整体宣传，扩大了西岸地区的社会影响力；2013 年西岸建筑与当代艺术双年展推动了徐汇滨江的文化发展和工业遗存的利用；2014 年 3 月 28 日龙美术馆西岸馆开幕、5 月 17 日余德耀美术馆开幕、9 月 25 日第一届西岸艺术与设计博览会开幕，同时西岸艺术中心、西岸艺术品保税仓库投入使用运营，可以说以 2014 年为爆发点，西岸进入了一个文化艺术加速发展的时期。接下来，从 2015 年上海摄影艺术中心开幕、丁乙工作室、乔空间、大舍建筑事务所等一批艺术家工作室、建筑师事务所、画廊等文化艺术机构陆续入驻西岸文化艺术示范区，到 2018 年油罐艺术中心开幕，在文化的硬件方面，西岸无疑已经成为上海文化品牌建设的重要品牌之一。

时空摆渡——西岸文化艺术季

在 2015 年第一届以"城市更新"为主题的上海城市空间艺术季在西岸成功举办之后，随着西岸区域内场馆内容的不断丰富、文化活动的逐渐多元，受众也从专业人员逐渐拓展到普通大众。如果说城市中的建筑是城市的骨骼，那么文化内容则是城市的血肉，二者合一的城市才是完整的、鲜活的城市，此时文化艺术的软件和城市空间竞争的软实力就显得尤为重要，到了这个阶段，西岸文化艺术实践进一步发展的重要"摆渡人"便是西岸文化艺术季。

西岸文化艺术季既是一个品牌，也是一个平台。它是由上海西岸发起，自 2016 年开始，在每年春夏和秋冬两季举办的区域性文化品牌，同时也是西岸"全球首展首秀首发新地标"的核心平台。它以"西岸计划"为宗旨，以"来西岸、去发现"为理念，联动西岸美术馆大道和西岸传媒港等文化、商业空间，每年推出展览、演出、论坛、市集、讲座等近百场文化活动，打造多元、开放的文化平台，倡导主动探索世界、发现自我的文化精神，致力于向世界传达中国文化的当代价值，打造上海文化品牌的金名片。

2017 年西岸文化艺术季持续 8 个月，举办文化活动百余场，除西岸主承办的西岸食尚

cannot be ignored.

Comprehensive promotional programs for West Bund Music Festival in 2012 expanded the social influences of West Bund; West Bund 2013: A Biennial of Architecture and Contemporary Art motivated the cultural development of industrial heritage utilization in Xuhui Riverside; and the year 2014 was "explosion point" after which West Bund entered an accelerating stage in terms of cultural and art development, with the opening of Long Museum (West Bund) at 28 March, Yuz Museum at 17 May and the first West Bund Art & Design at 25 September, and the completion of Art West Bund and West Bund Fine Art Storage. Starting from the opening of Shanghai Centre of Photography in 2015, a number of cultural-art institutions including Art Studio of Ding Yi, the gallery Qiao Space and Atelier Deshouse were settled in the West Bund Culture and Art Pilot Zone. By the opening of Tank Shanghai in 2018, West Bund was unquestionably an important cultural brand of Shanghai in terms of physical cultural infrastructure.

Ferry between Times and Spaces: Art' West Bund

After the success of SUSAS 2015 "Urban Regeneration" in West Bund, more cultural institutions and events appeared in this area, and the audience extended beyond professionals to the wider public. If buildings constitute the skeleton of a city, culture is the flesh, and every whole and living city must combine both. At this stage, the intangible elements of culture and art and the soft capacity of competitive urban spaces are particularly important, and the essential "ferry" to carry on the cultural and art practices in West Bund is Art' West Bund.

Art's West Bund is at once a brand and a platform. Since its initiation in 2016 by West Bund, the regional cultural brand has been offering two seasons – SS for Spring and Summer, and FW for fall and winter – every year. It is also the core stage for West Bund as "the New Landmark of Global Debut Destination" of West Bund. In a year, Art's West Bund, given the objectives of "West Bund Project" and the concept of "To Discover", offers nearly 100 cultural events – exhibitions, performances, forums, fairs and lectures – in collaboration with cultural or commercial spaces such as West Bund Museum Avenue and West Bund Media Port. It has also established an open and diversified cultural platform which encourages the spirit for exploring the world and the self, endeavors to communicate contemporary Chinese values to the world, and aims to be an exemplar cultural brand of Shanghai.

Art' West Bund 2017 presented over 100 events in 8 months which, in addition to West Bund Food Festival, West Bund Music Festival, West Bund Art & Design, the Stage West Bund and other events hosted by West Bund itself, included 25 invited medium-to-large market events and over 70 exhibitions organized by the cultural institutions in West Bund. And Art' West

西岸文化艺术季
Art' West Bund

节、西岸音乐节、西岸艺博会、西岸艺场等活动外，引进大中型市场活动25场，区域内文化机构自办展览70余场。同时，进一步整合、挖掘更多社会主体参与艺术季，例如2017年完全引入ONTIMESHOW时尚展会活动，新开了大田秀则画廊、阿拉里奥画廊、东画廊、油罐项目空间、SSSSTART五家艺术机构。

2018西岸文化艺术季继续鼓励公众参与，持续扩大群体及受众规模，同时也开始尝试以市场为主导的转型，让文化艺术事件与城市区域整体发展步调相匹配，确保可持续发展的可能性，而非仅仅为事件而事件。

到了2019年，西岸文化艺术季在春夏和秋冬两季联动西岸腹地20多个场馆，举办了100余场核心活动、120场系列活动。春夏季举办第十季ONTIMESHOW、teamLab上海首展、Chanel内地首展、Sneaker Con内地首展、BMW新车首发等多场艺术时尚与品牌首秀。秋冬季由世界人工智能大会·徐汇西岸会场的"双A"体验揭开帷幕，同时围绕首届上海国际艺术品交易月，共呈现30天30展120场活动。

可以明显看到，西岸文化艺术季经过几年的发展，在活动的丰富性和多元性上有了进一步强化，由原先以美术馆自办活动为主，逐步发展出涵盖艺术、时尚、潮流文化、品牌发布等多元跨界的复合属性。同时从活动类

Bund was further involving and seeking out new participants. For example, the fashion show ONTIMESHOW decided to permanently settle in West Bund in 2017. And five art institutions were opened in the same year: Ota Fine Arts, Arario Gallery, Don Gallery, Tank Shanghai and SSSSTART.

Art' West Bund 2018 continued to encourage public participation, trying to attract a larger and more diversified audience. At the same time, it began to experiment a market-led reorientation: synchronize cultural-art events with the overall developmental pace of Shanghai and guarantee sustainability of events, instead of "events for events' sake".

Art' West Bund 2019 – SS and FW - offered over 100 major events in collaboration with more than 20 venues in the hinterland of West Bund, and 120 serial activities. The program of SS season includes the 10th ONTIMESHOW and multiple debuts such as debut of teamLab in Shanghai, Chanel and Sneaker Con in Mainland China, and new BMW model. The FW season was opened by the "AA West Bund Pass" of World Artificial Intelligence Conference + Art' West Bund, and offered 30 exhibitions and 120 activities in a period of 30 days focusing on the 1st Shanghai International Artwork Trade Month.

Through its years of development, Art' West Bund has seen obvious progress in richness and diversity – instead of a program consisting largely of events hosted by West Bund Museum, it has gradually grown to one featuring hybridity and mixture of art, fashion, popular culture and commercial debut. And aiming to refresh and enrich the diversified local life experiences, Art' West Bund is devoted to exploring ways to break down the barriers between commercial, commercial and public spaces from the perspectives of event type, commercial influence and marketability.

型、市场影响力、市场化程度等维度，着力探索打通文化空间、商务空间和公共空间隔阂的可能性，以刷新和丰富区域多元化的生活体验。

空间实验——西岸智能导览系统

如果说人类发展的历史同时也是一部空间构造的历史，从远古洞穴到摩天大楼、从城市空间到赛博空间，今天的你我早已习惯在互相平行的城市空间和网络空间中构筑对生活的理想。如果说西岸"十年磨一剑"，聚焦规划、生态、文化、科创，努力打造的卓越水岸是在城市物理空间的有机更新。那么面对网络空间对城市空间的叠加与渗透，我们要如何基于已有的实践，借助网络和数据，构筑一个更为智慧的水岸，实现城市在物理空间和网络空间的有机连结和更新？

带着这样的发问，我们立足西岸十多年的开发建设，已建成的 8.95 公里沿江景观岸线、50 万平方米公共开放空间、龙美术馆、余德耀美术馆、油罐艺术中心等沿岸工业遗存改建的公共艺术空间、以西岸美术馆为代表的新建公共文化设施和一系列艺术实践，以"2019 上海城市空间艺术季实践案例展——上海西岸艺术实践"为契机，做了一次实验。

实践案例展通过搭建西岸智能导览系统，实现覆盖全区域定位导航，精细化至全区域艺术场馆及沿江配套服务设施点位；整合区域所有艺术场馆、公共文化设施、工业历史遗迹、沿江景观节点的信息，提供导览介绍服务；围绕西岸生活服务和活动推广，推送区域全面、实时的动态资讯的资讯服务。同时，展览基于西岸智能导览系统，推出"西岸空间艺术打卡路线"活动，吸引市民前往指定打卡地点寻找二维码扫码打卡，串联上海西岸艺术实践，让市民亲身体验近年来徐汇滨江的艺术实践案例，探索西岸发展背后的故事。打卡活动发挥智能导览、导航的互动功能优势，串联沿江 3 大艺术装置、7 大文化场馆、9 大工业地标，上线 19 天累计千余游客竞相参与打卡，成为上海城市空间艺术季徐汇案例展的重磅活动。

此外，西岸智能导览还结合票务服务系统，

Spatial Experiment: West Bund Smart Tour Guide System

If the history of humans is also one of spatial constructions – from prehistoric caves to skyscrapers, from urban spaces to cyber matrices – we have long been accustomed to structuring our life ideals in the parallel twin spaces of urbanity and Internet. The consistent decade-long efforts in West Bund, focusing on planning, ecology, culture and technological innovation and aiming to build an excellent waterfront, constitute an organic regeneration of the urban physical spaces. Then as the Internet spaces superimposing on and infiltrating into our city spaces, how can we, based on previous practices and with the forces of Internet and data, construct a smarter waterfront which enables an organically renewed connection between physical and virtual spaces?

To answer the question and seizing the opportunity of "SUSAS 2019 site project: Art Practices in West Bund, Shanghai", we have conducted an experiment. The experiment is based on the developments and constructions in the last decade: an 8.95km landscaped waterfront, public open spaces of 500,000m^2, riverside public art venues transformed from industrial heritages such as Long Museum, Yuz Museum and Tank Shanghai, new public cultural facilities such as West Bund Museum, and a series of art practices.

By building West Bund Smart Tour Guide System, the site project provides navigational services that cover the entire West Bund and are fine-grained enough to include all art destinations and waterfront amenities; integrates information about all art destinations, public cultural facilities, industrial heritages and waterfront scenic spots as a smart tour guide; and establishes a comprehensive and real-time dynamic information service focusing on daily life amenities and new events in West Bund. And based on the smart system, the West Bund Public Art Tour invites citizens to search for and scan QR codes at a set of check spots that connect art practices in West Bund, and in this process personally experience the recent art projects in Xuhui Riverside and explore the stories behind the development of West Bund. The Art Tour utilizes the interactive features of navigation and tour guide, and incorporates 3 art installments, 7 cultural venues and 9 industrial landmarks along Huangpu River. More than 1,000 visitors participated in the event during its 19 days of service, making it a keystone of SUSAS site project in Xuhui District.

Two convenient and affordable multiple-venue passes - AA Pass and West Bund Time Pass – are rolled out based on the WeChat service account of Art' West Bund and in collaboration with ticket service systems. Such passes integrate features including landmark tour guide, area navigation, ticket booking and newsletter, and are becoming the primary ticket platform of West Bund Museum instrumental for venue operation and big-data administration, as well as a helpful guide for life in West

西岸文化艺术季智能服务平台
Art' West Bund smart hub

依托西岸文化艺术季微信服务号，集地标导览、区域导航、票务预约、活动资讯等服务性功能开发于一体，尝试以"西岸双A体验"和"西岸巅峰艺术月"为主题，策划了两轮便民、惠民的多馆联票，并作为西岸美术馆的主流票务平台，在票务运营、大数据管理等方面发挥了重要作用，正在成为西岸城区生活指南和助手。2019 秋冬季第二轮多馆联票——"西岸巅峰艺术月"联票联动了西岸 6 馆 12 展，开票 5 天即全部售罄，成为上海整个 11 月最炙手可热的西岸艺术"一卡通"，为平台创下了单条阅读量近万，单日粉丝增长近 700 的数据高峰。

西岸多年来的实践和发展除了注重城市空间中社会与文化的实践，同时也重视发挥想象力和洞察力，在包容性增长和可持续发展的目标下，让现实空间与日益重要的网络空间深度融合，服务于人群的需求和使用。三届空间艺术季成为西岸展现"铁锈地带"到迈向卓越水岸的舞台。希望"西岸时刻"的"西岸经验"不但属于西岸，而且为今后城市空间的发展提供方向，让城市和文化的演绎、科技和艺术的碰撞流动在城市空间的发展中。

Bund. The second pass in Art' West Bund FW 2019, West Bund Time Pass, involves 6 venues and 12 exhibitions in West Bund. Sold out in merely five days, it is the most wanted "admissions-in-one" of art in the November of West Bund, and contributes to a statistical spike: the article about the pass was read by nearly 10,000 persons and almost 700 new followers flooded into the WeChat account of West Bund.

In its decade of practice and development, West Bund not only emphasizes social and cultural practices in urban spaces, but focuses on, given the goals of inclusive growth and sustainable development, deeply integrating physical spaces with the increasingly important cyber spaces through the power of imagination and insight, catering to demands and usages of the people. SUSAS, as from 2015 to 2019, have provided a stage for West Bund to show its transformation from Rut Belt to Excellent Waterfront. We hope that the moments and experiences of West Bund are not limited to itself, but may provide directions for developing urban spaces in the future, engaging cities and cultures in an interactive and dynamic process of performance, technology and art.

新城建设回顾与
滨水空间微更新

Review of New City
Construction and Waterfront
Micro-renewal Projects

5000 年的相遇，当广富林遇到 2035——松江新城实践案例展

Site Project in Songjiang New City — Five Thousand Years between Guangfulin Heritage and 2035

金英、翟伟琴
松江区规划和自然资源局

Jin Ying, Zhai Weiqin
Songjiang District Planning and Natural Resources Bureau

主办单位：
上海市松江区人民政府
承办单位：
松江区规划和自然资源局、松江区文化和
旅游局、松江区体育局、松江新城建设发
展有限公司
策展人：
翟伟琴
展览时间：
2019 年 9 月—2019 年 12 月
展览地点：
松江区城市规划展示馆

Host:
People's Government of Songjiang District, Shanghai
Organizers:
Songjiang District Planning and Natural Resources Bureau,
Songjiang Administration of Culture and Tourism, Songjiang
Administration of Sports, Songjiang Xincheng Construction
Development Co., Ltd.
Curator:
Zhai Weiqin
Time:
September 2019 – December 2019
Venue:
Songjiang Urban Planning Exhibition Hall

松江，以水为名，遇水而兴。水为城之血脉，孕育滋养了这片富饶的土地，也见证了曲水流觞和大江奔腾中的城市变迁。

2019 年 9 月—2019 年 12 月，在各级领导的支持和相关单位的通力协作下，伴随着"2019 上海城市空间艺术季"的滨水空间主题，松江区在城市规划展示馆开启了"5000 年的相遇，当广富林遇到 2035"的实践案例系列活动，表达了"滨水空间为人们带来美好生活"的寓意，吸引公众参与沉浸其中，在这里与城市的过去、现在和未来相遇。

此次松江案例展主要由风貌影像展和各项市民活动组成。通过影像、图片和文字，用公共空间的艺术语言，讲述了松江千百年来的城市和人文历史，通过沉浸式展示体验，如同徜徉在历史文化画卷中……

通过松江新城的发展演变历程和十里华亭湖的实践案例展示，呈现了不同时期的历史遗

Songjiang is named after and thrives on the river of Wusong. Water flows through the veins of Songjiang, nurturing this fertile land and witnessing its transformation among winding and thundering streams.

The site project "Five Thousand Years between Guangfulin Heritage and 2035", with the support of governmental leaders and collaboration of involved agencies, was held between September and December 2019 at Songjiang Urban Planning Exhibition Hall. Echoing the waterfront theme of SUSAS 2019, it expresses the meaning of "how waterfronts bring wonderful life to people" and invites citizens to participate and get themselves immersed in the site project, encountering the past, present and future of city here.

The Songjiang site project consists of an exhibition of scenic images and a collection of public activities. It tells the stories of Songjiang through its millennials of history in the forms of photography, imagery and text, and using the artistic language of public spaces and immersive showcasing experiences, invites visitors to walk into a scroll of culture and history...

By demonstrating the developmental process of Songjiang New City and the sceneries of Huating Lake distributed among a belt of 10 li (5km), the site project shows us the historical relics

"5000 年的相遇，当广富林遇到 2035"主视觉海报
Key visual of "Five Thousand Years between Guangfulin Heritage and 2035"

迹、风貌留存和人文胜景，体现了古今城市与滨水空间的完美融合。

松江城市建设的再腾飞，源于 20 世纪 90 年代中后期。规划方案几经论证，提出了规划建设新城区的构想，确定以老城区为基础，规划发展沪杭高速北侧的新城区，并先期建设新城示范区。

随着 21 世纪上海"十五"期间在郊区重点建设"一城九镇"的决定出台，确立了松江新城作为"一城"的发展构想。

一座城市如同一件作品，只有精心打磨才能完美呈现。随着新一轮《上海市城市总体规划（2017—2035 年）》的提出，要建设卓越的全球城市，成为令人向往的创新之城、人文之城、生态之城，规划思路、理念和方法都需有很大转变和提升。在新的目标背景下，战略地位的转变，对松江的未来发展提出了新的挑战和机遇。松江区总体规划（2017—2035 年）的确立，既是对松江可持续发展的进一步思考和全面探索，也将通过规划编制并稳步推进、助力建设"科创、人文、生态"的现代化新松江，将形成"一廊一轴，五带四片"的空间结构，以更具综合性、独立性的长三角节点城市的"新定位"，来支撑上海打造全球城市，参与全球竞争。

滨水空间见证人类文明发展的历史，如今更

and cultural landscapes in various periods of Songjiang, and embodies a perfect fusion of ancient town, modern city and waterfront spaces.

The redevelopment plan of Songjiang district can be traced to the latter half of 1990s. After multiple rounds of research and evaluation, the vision of Songjiang New City was finalized: based on the historical quarters of the town, new developments were planned to be on the north of Shanghai – Hangzhou Highway, beginning with a demonstrative zone of new city.

When it came to the 21st century, Songjiang New City was designated as the "One City" of the prioritized developments in suburban Shanghai — "One City and Nine Towns" — during the period of 10th Five-Year Plan.

A city is like a craftwork whose perfection is only possible with careful polishing. The new *Shanghai Master Plan (2017 – 2035)* has called for radically changed and improved approaches, concepts and methods of urban planning to build an excellent global city, an admirable city of innovation, humanity and ecology. Facing new goals and shifts in strategical status, Songjiang is met with new challenges and opportunities in its future development. Songjiang District Master Plan (2017 – 2035) is an extended reflection and comprehensive exploration about the sustainability of Songjiang, aiming to plan for and steadily promote and support the development of a modernized city featuring "innovation, humanity and ecology", establish a spatial structure of "One Corridor, One Axis, Five Belts and Four Zones". Given the "new positioning" as a more comprehensive and independent node among the cities of Yangtze Delta, Songjiang will support Shanghai in its efforts to build a global city and participate in global-scale competition.

Waterfronts have witnessed the history of human civilization, and

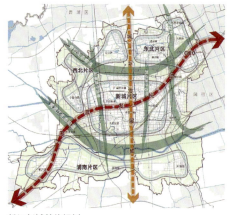

松江新城总体规划
Master plan of Songjiang New City

now serve as meeting and leisure spaces for citizens. According to Songjiang District Master Plan (2017 – 2035), the banks of Huating Lake will be a new culture and leisure landmark of Songjiang.

The 10 li area up and down Huating Lake, with its millennial history, connects the four essential parts of Songjiang in its historic transformation: Guangfulin Heritage of ancient human civilization dated 5,000 years ago; the historical town of Cangcheng in the dynasties of Ming and Qing; Thames Town representing the landscape of Shanghai in the era of "One City and Nine Towns"; and the academic and culture destination, Sonjiang University Town. The site project demonstrates the

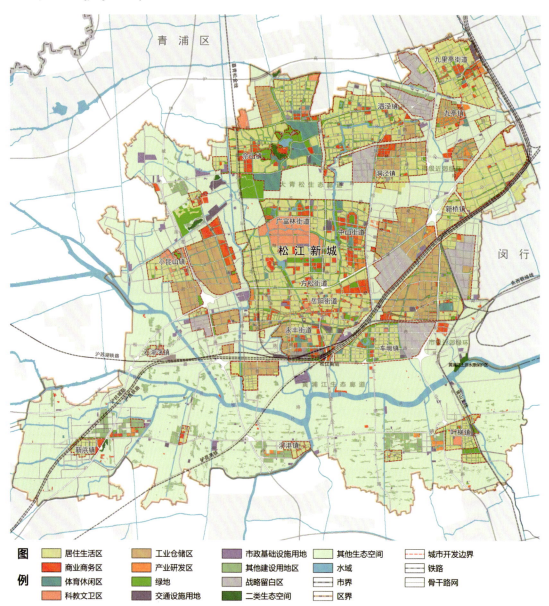

图	居住生活区	工业仓储区	市政基础设施用地	其他生态空间	- - - 城市开发边界
例	商业商务区	产业研发区	其他建设地区	水域	铁路
	体育休闲区	绿地	战略留白区	市界	骨干路网
	科教文卫区	交通设施用地	二类生态空间	区界	

松江新城总体规划
Master plan of Songjiang New City

仓城
Cangcheng

是市民休闲聚集的场所。松江区总体规划
（2017—2035 年）确定，华亭湖两侧的滨
水空间将成为松江文化休闲的新标杆。

穿越千年的华亭湖上下游十里串联起松江古
今华丽变身最重要的四大板块：从 5000 年
前的广富林遗址文化到明清时期仓城历史风
貌；从上海"一城九镇"时期的泰晤士风貌
小镇到充满学术气氛的大学城人文胜景等。
此次实践案例展对不同时期华亭湖两岸的风
貌展示，如串起的颗颗明珠璀璨闪耀，吸引
人们驻足欣赏，仿佛穿越了时空隧道……

"寻上海之根，游华亭十里"，作为华亭湖
滨河空间的深度体验项目规划，也将开启上
海城市沿河滨江旅游的新篇章，打造"华亭
十里水上游"的新品牌，融合全域旅游概念
提供优质旅游服务，引导优化滨江两岸空间
成为新一轮城市更新典范。

以滨水空间为特色的华亭湖寻根之旅，融合
了古今历史，也是一次文化之旅。

仓城风貌区是古市河向北连接华亭湖的起
点，在沿河两岸绵延扩展的城池，规模甚大，
有城墙及四门、敌楼、瓮城等，明清时期达
到鼎盛，是中国古代最大的国粮地和漕运始
发地。漕运船队由仓城出发，经京杭大运河
将松江大米、棉布等重要物资经古市河发通
达京城。仓城依河兴筑，因河兴盛，一河一
街一屋城市肌理，是风貌区滨水空间的灵魂
所在，积淀了深厚的文化底蕴。

"华亭十里水上游"项目
Huating Lake Cruise

sceneries by Huating Lake in various times, inviting people to stop and appreciate these connected shining jewels of Songjiang as if travelling through a time tunnel ...

The plan for experiential programs along Huating Lake – Visit Huating Lake and Search for the Roots of Shanghai – will open a new chapter of waterfront tourism in Shanghai and create the new brand of "Huating Lake Cruise". It intends to provide premium services incorporating the concept of holistic tourism, guiding and optimizing the Huating waterfronts to be an example for recent urban regenerations.

The waterfront-themed and root-searching trip of Huating Lake is a cultural experience that brings ancient and contemporary histories together.

The Cangcheng Scenic Zone is where its Town River flows northward into Huating Lake. Cangcheng ("warehouse town") is a massive fortified town stretching along the river with walls, four gates, watchtowers and barbicans. Cangcheng reached its most prosperous moments during the dynasties of Ming and Qing when it was the largest state-owned grain depot throughout the entire history of ancient China, and the starting point of Caoyun, or transporting produces of South China to the imperial capital. All Caoyun fleets carrying important goods of Songjiang such

泰晤士小镇钟书阁
Zhongshuge Bookstore in Thames Town

华东政法大学
East China University of Political Science and Law

暖阳映红广富林
Sun rays on Guangfulin Heritage Park

松江新城产业发展总体布局示意图
Industrial development diagram of Songjiang New City

华亭湖沿岸的泰晤士小镇，是松江新城区重点建设的核心区域，占地面积 1 平方公里，总建筑面积 50 万平方米，小镇水网密布，既借鉴了英国泰晤士河沿岸的建筑风貌，也结合了中国人实际生活习惯，从整体布局到一砖一瓦既体现了原汁原味的英伦风情，又体现了松江新城浓烈的现代化、国际性、生态型以及旅游文化气息，滨水建筑空间在这里完美呈现。

松江自古多学府，人文艺术成就辉煌。与松江新城同步建设的上海松江大学城，有上海外国语大学、上海对外经贸大学、上海立信会计金融学院、东华大学、上海工程技术大学、华东政法大学、上海视觉艺术学院七所高校，以环境园林化、布局科学化、设施现代化、管理智能化而声名远播，建筑风格鲜明且有特色。各校区滨水而建，气势恢宏、环境优美，是莘莘学子向往的学术殿堂。

千年上海看松江，千年文明在松江。早在 5000 年前的新石器时代，就有祖先在广富林一带依山傍水繁衍生息、渔猎耕作，用水火和土石开荒拓地，创造了"广富林文化"，上海的根系由此繁密起来，上海古文明序幕自此拉开。远古文化留下的珍贵的遗存，是优秀传统文化的重要传承地，是以上海历史文脉为主线的古文化遗址保护和旅游目的地，在生态、人文上体现高品质。

as rice and cotton cloth used to sail from Cangcheng through its Town River, and then through the Grand Canal until their destination, Peking. Cangcheng was constructed by and thrived on its river. The river, the main street and the historical houses constitute the culturally enriched fabric and soul of Cangcheng Scenic Zone.

Thames Town on the bank of Huating Lake is the primary development of Songjiang New City which covers an area of 1km². Its total building area is 500,000m². This project, with its dense water network, borrows from the buildings along Thames in the United Kingdom and fits well into the habits of Chinese people. From the master plan to its every brick and tile, Thames Town reflects an authentic British flavor and embodies the modern, international, ecological and tourist features of Songjiang New City. Here is a perfect demonstration of waterfront architecture.

Songjiang has a history of academies and intellectual accomplishments. Songjiang University Town which was developed at the same time with Songjiang New City, now contains seven higher education institutions: Shanghai International Studies University, Shanghai University of International Business and Economics, Shanghai Lixin University of Commerce, Donghua University, Shanghai University of Engineering Sciences, East China University of Political Science and Law, and Shanghai Institute of Visual Art. The University Town is renowned for its garden-like environment, well-considered layout, modern facilities and smart administration, and its buildings are characteristically designed. These waterfront institutions are massive yet elegant academies desired by Chinese students.

Songjiang, Shanghai has a civilization history of thousands of years. Five thousand years ago, our Neolithic ancestors are already found to dwell and multiply, fish and hunt, opened up wastelands with fire and stoneware, grow crops in the region of Guangfulin. The Guangfulin Culture marks the beginning of ancient civilization and the roots of what is now known as Shanghai. The precious ancient relics constitute an important heritage site of the brilliant cultural history of Shanghai, and a protected site and tourist destination of ancient relics that carries through the cultural vein of Shanghai. It reflects splendid ecological and humane values.

"Brick, tile and old lanes, corridor, hub and new city" – Songjiang is now embracing a new opportunity for development after its history of 5,000 years, with Songjiang New City, one of the five New Cities enlisted in the developmental strategies of Shanghai, standing at the new starting point delineated in the 14th Five-year Plan. The five New Cities, including Songjiang New City, will put forth a new wave of planning and construction. Focusing on the positioning of "independent and comprehensive city", they are intended to be cities of future as pilots for high-quality

松江新城鸟瞰
Bird view of Songjiang New City

"一砖一瓦一古巷、一廊一纽一新城"。历经 5000 年后的今天，站在"十四五"发展新起点，被列入上海"五个新城"发展战略的松江新城，迎来了新的发展契机。松江等五大新城将努力推进新一轮新城规划建设，聚焦"独立的综合性城市定位"，建设成为引领高品质生活的未来之城，要建成现代化的大城市和长三角的增长极，推进城市建设的创新实践区、数字化转型的示范区和上海服务辐射长三角的战略支撑点，将培育成为长三角城市群中具有辐射带动能力的综合性节点城市。松江要建设成为上海在全球卓越城市的西南门户，建设科创松江、生态松江、人文松江，对打造生态空间和人文传承保护提出了更高目标与要求。

人居环境品质不断提升和优化，形成优于中心城的蓝绿交织、开放贯通的"大生态"格局，骨干河道两侧和主要湖泊周边实现公共空间贯通，率先确立绿色低碳、智慧数字、安全韧性的空间治理新模式。

让我们共同努力和期待一个精心规划和打磨的"古今融合"的新松江，在不久将来的完美呈现！

简·雅各布斯说："文明的价值就在于让生活方式更加复杂，因为更复杂、更深入的思

life, modernized mega cities as the growth pole of Yangtze Delta, spaces for creative practices of urban construction, demonstrative zones for digital transformation and strategic anchors for Shanghai to support the development of Yangtze Delta in general. And they aim to be comprehensive hub cities of strong catalytic roles in the region of Yangtze Delta. Songjiang will be built to be the southwestern gateway of Shanghai, the excellent global city, and a city featuring technological innovation, ecology and culture. Therefore, higher aims and requirements are imposed on the ecological environment and cultural inheritance of Songjiang.

Songjiang will keep improving and optimizing its living environment so as to establish an open and connected pattern of "Macrohabitat", with its fabric of greens and waters, that is superior to the downtown area. Public spaces along its main watercourse and around its bigger lakes will be opened, and Songjiang will firstly adopt the new mode of urban space management featuring environmentalism, low-carbon economy, smart technology, safety and resilience.

Let's work together and expect for a finely planned and polished New Songjiang that incorporates history with contemporary in the near future!

As Jane Jacobs quotes Oliver Wendell Holmes, Jr, "...the chief worth of civilization is just that it makes the means of living more complex...Because more complex and intense intellectual efforts mean a fuller and richer life. They mean more life." Indeed, besides making sure that the crowd is fed, clothed, sheltered and moved from place to place, cities also make life richer by satisfying higher spiritual demands, that is,

考和体验意味着更充实、更丰富的美好生活，意味着更旺盛的生命。"的确，城市让生活更美好的方式，除了解决衣食住行等必要的物质要求，还应有更高的精神需求，就是满足人们对更美好城市生活的向往。

上海城市空间艺术季活动以"文化兴市、艺术建城"为理念，通过城市滨河空间的展览演绎体验、公共艺术介入、交流合作思考等多种形式，完整呈现滨水空间更新的实践成果，有力推动了历史人文和现代理念的融合发展，着力打造出具有 " 国际性、公众性、实践性"的城市空间艺术品牌。

据不完全统计，松江 2019 城市艺术季实践案例观展和参加各种活动人数达到几十万人。进入城市艺术季展区，人们如同穿越到了城市过去，对话城市历史的同时，也感受到了未来城市的美好发展，对建设全球化的美好城市已经充满期待。

千年城市沧桑巨变，溯河追源、传承文明、把握机遇、与时俱进，是我们这代人的使命和责任，通过对美好城市生活的不懈努力和追求，我们相信未来一定可期……

people's desire for a better urban life.

Adhering to the concept of "Culture Enriches City, Art Enlightens Space", SUSAS is designed to establish a brand in urban space art events that features "internationality, publicity, and practicality" in the forms of exhibition, performance, public art, dialogue and collaborative program. It offers a comprehensive demonstration of the practical accomplishments as we regenerate waterfronts, and a powerful motivation for the integrated development of historical values and modern concepts.

According to an underestimated statistic, hundreds of thousands of people visited the SUSAS 2019 site project and other activities in Songjiang District. Entering the site project is like travelling back to the past of Songjiang, dialoguing with its history while appreciating its wonderful future and expected to a beautiful global city.

It is the mission and responsibility of our generation to trace the roots of Songjiang, a city which has undergone fundamental changes through its history of thousands of years, inherit its culture, seize new opportunities and keep paces with the times. We believe in its promising future if we incessantly pursue wonderful urban lives.

上海之南——奉贤新城实践案例展

The Shanghai South — Site Project in Fengxian New City

冯路
英国谢菲尔德大学建筑学博士
无样建筑工作室主持建筑师

Feng Lu
Architectural PhD of University of Sheffield
Chief Architect of Wuyang Architecture

主办单位：
上海市奉贤区人民政府
承办单位：
奉贤区规划和自然资源局
协办单位：
上海奉贤南桥新城建设发展有限公司
策展人：
冯路
展览时间：
2019 年 11 月 15 日—2020 年 2 月
展览地点：
奉贤区上海之鱼湖畔贤庐

Host:
People's Government of Fengxian District, Shanghai
Organizer:
Fengxian District Planning and Natural Resources Bureau
Co-organizer:
Shanghai Fengxian Nanqiao New City Construction and
Development Limited Company
Curator:
Feng Lu
Time:
15 November 2019 – February 2020
Venue:
Xianlu by Shanghai Fish Lake

"上海之南"是 2019 年上海城市空间艺术季在奉贤新城的实践案例展。展览从两方面呼应了本届空间艺术季所提出的"滨水空间为人类带来美好生活"之主题。一方面在于展览的内容，对公众介绍了奉贤新城过去十年建设发展中的诸多规划和建筑设计案例，其中有不少案例都与滨水空间有关。另一方面，展览本身作为金海湖畔的一个公共活动，它给展场所在地带来了新的变化，因而成为一个与空间艺术季主题相对应的空间实践。

奉贤位于上海的正南方。奉贤新城的前身是南桥新城。作为奉贤城区起点的南桥镇在地理文化上属于历史悠长的江南水乡区域，而与此同时，奉贤新城作为国际大都市上海的一部分，在城市发展新时期又必将承载着带有海洋气息的"南方"这一特定概念所内含的改革和创新精神。因此，本次展览拟定主题名称为"上海之南"，以表达奉贤新城所内含的文化特质，在历史与当下、继承与创

The SUSAS 2019 site project "The Shanghai South" at Fengxian New City echoes the theme "How Waterfronts Bring Wonderful Life to People" in two ways. First, among the planning cases and architectural designs of Fengxian New City in the last decade that are introduced in the site project, quite some are related to waterfronts. Second, as a public event by Jinhai Lake, also known as Shanghai Fish Lake, it changes the venue in new ways and hence constitutes a spatial practice in alignment with the theme of SUSAS 2019.

Fengxian District is the southern part of Shanghai, and Fengxian New City used to be called Nanqiao New City. Nanqiao Town at the rim of urban areas of Fengxian District is, in terms of geography and culture, a traditional Jiangnan water town while in this new era of city development, Fengxian New City, as part of the metropolitan Shanghai, has to carry the reformative and creative spirits inherent in the oceanic notion of "South" – hence "The Shanghai South". The theme expresses the inherent cultural characteristics of Fengxian New City, rescans the cultural identity, positioning and mission of New City projects from the perspectives of past and present, inheritance and innovation, and envisions the development of Fengxian as it shifts from a "satellite city" towards a "node city".

"上海之南"展览主海报
Poster of the Shanghai South

新的思考之中重新审视新城建设的文化身份、定位和使命，展望从"郊区新城"模式向"节点城市"模式转变过程中的城市发展。

作为上海市重点推进建设的新城之一，奉贤新城已经开发建设了十年。2008年1月上海市奉贤南桥新城建设发展有限公司成立，南桥新城开发管理委员会也于2月成立。同年，上海市政府批复了《上海市南桥新城总体规划（2004—2020年）》。南桥新城从此翻开了自己的篇章。之后伴随着上海市城市总体规划的调整，2017年南桥新城更名为奉贤新城。在2017年发布的《上海市城市总体规划（2017—2035年）》中，奉贤

Fengxian New City, as a major new city development of Shanghai, has been constructed for ten years. Fengxian Nanqiao New City Construction and Development Limited Company was founded in January 2008, and Management Committee of Nanqiao Xincheng Development in February. In the same year, the Municipal Government of Shanghai approved *Shanghai Nanqiao New City Master Plan (2004 – 2020)*. Since then, Nanqiao New City has opened a new chapter. In 2017, accompanying the changes in Shanghai Master Plan, Nanqiao New City was renamed as Fengxian New City. According to *Shanghai Master Plan (2017 – 2035)*, Fengxian New City, as one of the five prioritized new cities in Shanghai, is positioned as "a node city of Coast Corridor and Riverside Corridor, and a comprehensive service-based core city through which the northern bank of Hangzhou Bay influences Yangtze Delta" and therefore receives even more significant tasks. Given the rapid urbanization in

新城作为上海市重点推进的五大新城之一，被定位为"滨江沿海发展廊道上的节点城市、杭州湾北岸辐射长三角的综合性服务型核心城市"，使得新城被赋予了更加重要的使命。面对中国当代高速城市化进程，及时在城市建设过程中展开阶段性记录、思考和展望显得尤为重要。因此，展览选取了新城建设过程中的若干重要项目以及城市规划设计成果，回顾过去十年，并借此展望未来，期盼新城建设带来更美好的城市生活。

奉贤新城开发建设的前五年主要在于土地收储、规划调整、基础设施建设等工作，后五年则开始集中于重点城市区域和公共建筑的打造。滨水空间的塑造，始终伴随着新城建设而展开。从展览中所展示的规划和建筑设计中，我们不仅可以看到奉贤区 2035 总体规划从环杭州湾的大地理空间角度出发的规划布局，也可以看到像以金海湖为中心的新城核心区城市设计、金汇港生态滨水长廊规划、浦南运河金汇港"十字水街"规划、以及与浦南运河相互依存的南桥源城市复兴项目等具体的城市滨水空间规划设计研究与提案。其中，金海湖公园从 2010 年湖面一期工程开工，到 2018 年初步建成开放。2017年启动"十字水街"建设，以浦南运河北岸景观改造工程的完工为前期成果。这些滨水空间已经逐步成为奉贤新城内重要的公共空间。此外，新城邀请知名国内外建筑师设计的公共建筑，有一些也参与到滨水空间的塑造之中。例如日本建筑师藤本壮介设计的奉贤城市博物馆就坐落于金海湖畔，它通过半室外公共空间的营造，使得建筑与自然景观叠加在一起，为金海湖公园增添了活力和吸引力。

城市公共建筑与公共空间的建成，提供了让人们与艺术、与生活、与历史相遇的场所。然而，城市建设的过程却是另一种"相遇"，其中有个人与记忆、个人与地方、不同工作岗位的人们之间的相遇，也有宏观的社会政治经济文化进程与具体地方、具体项目、具体个人的相遇。展览为此专门拍摄了一部以采访个人为主的纪录片作为展品与规划设计成果共同呈现。我为纪录片取名为《我们的

China, timely recording, reflecting on and envisioning the progress of a city through its stages of construction is particularly important. So a few key projects and urban planning/designs are selected for the SUSAS site project as a way to review the last decade and imagine the future of Fengxian New City, in the hope that urban construction will bring a better city life.

The first five years of Fengxian New City development focus on land banking, planning adjustment and infrastructure construction, and the latter five years emphasize key urban areas and public buildings. The shaping of waterfronts follows the entire construction process of Fengxian New City. The exhibited urban planning and architectural designs not only demonstrate the large-scale spatial layout around Hangzhou Bay in accordance with "Shanghai Master Plan 2035", but also the design for the core area of Fengxian New City with Jinhai Lake at its centre, the plan for the ecological waterfront corridor of Jinhui Harbor, the plan for "Cross Water Streets" at Punan Canal and Jinhui Harbor, the Nanqiaoyuan Regeneration Program interdependent with Punan Canal and other specific plans, designs, researches and proposals concerning urban waterfronts. For example, Jinhai Lake Park began the first phase of water-surface project in 2010 and was basically completed by 2018; "Cross Water Streets" was initiated in 2017 and its first accomplishment is marked by the landscape transformation project of the northern bank of Punan Canal. These waterfronts have gradually become essential public spaces in Fengxian New City. What's more, some of the public buildings designed by Chinese or international renowned architects are also engaged in the shaping of waterfronts. For example, Fengxian City Museum designed by Japanese architect Sou Fujimoto, sits by Jinhai Lake and imbricate the building with natural landscapes with its half-open public spaces, adding to the vitality and attractiveness of Jinhai Lake Park.

Public buildings and public spaces in cities provide places where people encounter art, life and history. Yet there is another "encounter" in the process of urban construction: individuals meet memories and places, people in different professions meet each other, macro social, political, economic and cultural processes meet specific places, projects and persons. So a documentary consisting primarily of individual interviews was produced as an item of the site project and exhibited together with works of urban planning and design. I call the documentary *Our New City* because I want to highlight that behind social development and urban renewal on the macro level, there are actually existing and participating individuals. The documentary was initially planned to be a 15-min clip, but eventually expanded to a 20-min one because we found many materials too valuable to abandon. The primary interviewees are eight persons related to Fengxian New City, including four local residents, three architects and an employee of Fengxian New City Construction and Development Company. Cai Xihao,

"上海之南"展场
Venue of the Shanghai South

金海湖城市公园
Jinhai Lake Urban Park

新城》，是为了强调宏观的社会发展和城市空间变更的背后，实际上都是诸多个体的存在和参与。纪录片的时长原本预定为 15 分钟，但后来因为内容素材丰富而无法舍弃，最终版本为 20 分钟。影片主要采访了 8 位与新城有关的个体，其中包括 4 位本地居民、3 位建筑师和 1 位新城公司职员。1941 年出生于奉贤的本地居民蔡锡豪，作为一个老城的居民，讲述了他记忆中奉贤的城市化进程与个人生活的变化。而另一位年长的本地居民屠荣，经历着原本居住的乡村转变为金海湖城市公园的时空巨变，动迁到新居之后，他与妻子依然会骑车到新城内空闲的土地上开垦菜地。对他的采访呈现了与城市的宏观变化相交织的普通人日常关心的事物和生活。汪胜嘉和韩东辰是在新城生活的年轻夫妇，他们曾经是金海湖畔新建住宅小区里最早一批的居民。他们讲述了在年轻人视角下对于新城生活的印象和期待。在本地居民之外，我们选取了在新城有建成作品的三位建筑师。他们既用外来者的眼光看待奉贤新城，又是新城建设实际的参与者。法国建筑师何斐德眼中的新城充满绿色与阳光，他设计的九棵树未来艺术中心于 2019 年建成开放。这个建筑物距离金海湖公园不远，功能包括

born in 1941, lives in the old town. He tells the urbanization of Fengxian as he remembers and the changes of his personal life in this process. Another elder resident Tu Rong has experienced fundamental changes as his home village was transformed into Jinhai Lake Park. After moving into their new home, the Tu couple still ride to vacant lands in New City and plant vegetables there. Interview with him shows how the daily life and happenings – what concerns common people – entangle with massive changes of the city. Wang Shengjia and Han Dongchen are a young couple living in Fengxian New City who used to be one of the first to move into the new apartment buildings by Jinhai Lake. They tell the impressions and expectations of life in Fengxian New City from the perspective of young people. Besides these local residents, we also invite three architects of some completed projects in Fengxian New City. They view the New City with eyes of outsiders, and actually take part in its construction. Frédéric Rolland, a French architect, sees Fengxian New city as a place of green and sunshine. His work, Nine Trees Future Art Centre, was completed in 2019. The facility not far from Jinhai Lake Park consists of three stages in various sizes. As a new urban landmark, the public building has obviously added a new cultural emblem to Fengxian. Senior-citizen University of Fengxian District, a new service building of multiple functions in the old town, provides elders in Fengxian a new kind of life. Its designer, Zhang Jiajing, not only introduces the design, but shares his opinions about the characteristics of Fengxian as a city. The third interviewed architect is Chen Jiawei whose designs are closely related to the local community – Jinhai Community Healthcare Centre is challenged by the relationship between a new structure and an old neighborhood

三个不同大小的演出剧场。作为新落成的城市标志性公共建筑，它显然给奉贤增加了新的文化符号。奉贤区老年大学是建于老城区内的新建筑，这个功能齐备的综合服务设施给奉贤的老年居民带来了不一样的生活。这个项目的建筑师张佳晶不仅介绍了设计本身，也分享了对于奉贤城市特质的个人看法。另一位建筑师陈嘉炜，他的设计作品与本地社区紧密相连。无论是金海社区卫生服务中心项目面临着新建筑与老社区的关系处理，还是金水苑社区生活驿站项目直面着新城老居民的日常生活，其背后都包含着城市、建筑与个人之间复杂而丰富的关联。3位建筑师在奉贤新城的设计作品也都是本次展览所展示的建筑项目，因而展示案例与纪录片之间建立了一种场内的对话。最后一位受访者是奉贤新城公司工程部的平燕女士。对于新城来说，她既是一位本地居民，又是一位建设者。作为新城公司最早的员工之一，她讲述了个人视角之下新城发展历程中遇到的困难和希望。在当代中国城市高速发展之下的每一个变化的背后，都必然包含着不同层面的、看不见的挑战和博弈，包含着多样化的个人记忆与经验，也包含着历史、现在与未来。就像纪录片的开场，工人在金海湖公园中吊装种植大树，这个缓慢的长镜头展现了新的生命周期的开始，而它也意味着时间、空间与场所的更替转变。

"上海之鱼"是金海湖城市公园的别名，因其以金鱼图案作为景观总平面而得名。它是奉贤新城核心的城市公共空间之一。伴随过去十年的新城建设，"上海之鱼"已经初显风貌。不仅城市公园已经基本建成，环湖游览步道也基本完成并开放使用。为了让更多的市民了解和体验到这一滨水空间的形成和魅力，展览以北侧湖畔的贤庐作为展览场地，一方面希望将专业展览与滨水空间的公共活动体验相结合，另一方面也希望借展览之契机给贤庐的场所状态带来新的可能性。

贤庐是一栋两层建筑，坐落在金海湖边。几年前建成的时候，其就是用作新城规划展示空间，后来因为各种原因，闲置了一段时间。它所在的场地是一片相对独立的、伸入湖中

and Jinshuiyuan Community Life Centre directly deals with the daily life of elder citizens in Fengxian New City – and underlies the sophisticated connections between city, building and human. All works of these three architects are also exhibited in the site project and hence initiates an internal dialogue between the documentary and the documented projects. The last interviewed person is Ping Yan, a member of the Engineering Department of Fengxian New City Construction and Development Company. She is at once a resident and a builder of Fengxian New City. As one of the earliest employees of the Company, she describes the difficulties and hopes in the development of Fengxian New City from her private perspective. Behind every change in the rapidly developing Chinese cities, there must be invisible challenges and gaming on various levels – there are diversified personal memories and experiences, and history, present and future are all concerned. The documentary starts with a slow long take of laborers planting a tree with hoist which represents the beginning of a new lifecycle, and the shifts in time, dimension and space.

Shanghai Fish Park, also known as Jinhai Lake Urban Park, is named after its master plan that resembles a goldfish. It is one of the central urban public spaces of Fengxian New City, and with the new city' development in the last decade, it has begun to show signs of charm. Not only the park, but also the promenade around the lake are basically completed and opened to the public. To allow more citizens to learn about and experience how this waterfront is formed and how appealing it is, the site project is located at Xianlu on the northern bank of the lake, in the hope of integrating professional exhibitions and public waterfront activities, and on the other hand seizing the opportunities of site project and thus bringing new possibilities to Xianlu.

Xianlu is a two-storey building by Jinhai Lake. It was used as an exhibition hall for the planning of Fengxian New City when it was completed several years ago, and was then left unused for some time. Now it is standing on a relatively independent triangular lawn that reaches into the lake. Since the building is unused, this waterfront whose plank road winding along the bank making it a rare landscape stroll path by the lake is also closed. Now that the site project is held at Xianlu, the building is reused. We have renovated its indoor spaces in response to the demands of exhibition, and in order to create more possible usages in the future. The modest changes made to the building opens up a closed space that invites the sceneries outside in. The exhibition room on the right hand side of the entrance hall used to be a closed place – the exhibitive wall blocked the French windows and kept out the sunshine and the sight of trees – and after we reopened it up, it is now a bright and spacious room. What's more, we have replaced the narrow steps through which this room led to the lake-facing main exhibition hall with a wider stairway and thus connects the room and the hall. The conference room in the centre of Xianlu is repurposed for playing the documentary, and the exhibited items, planning and design,

金海湖城市公园
Jinhai Lake Urban Park

的三角形绿地。因为建筑的闲置，这片滨水空间也处于封闭的状况。场地里面原有沿着水岸伸展的栈道，是湖边少有的亲水景观步道。展览以贤庐作为展场，使这个闲置的建筑又重新被使用。为了配合布展的空间要求，也为了给将来使用带来更多的可能性，我们重新翻新了室内空间。对于空间的改动并不多，却让原本相对封闭的室内空间变成一个自由流动的开放空间，同时也使得建筑外部的景观更好地进入建筑内部。建筑入口门厅右侧的展厅原来是一个封闭的房间，展墙挡住了落地窗也遮住了窗外的阳光和树木。我们把它重新打开变成一个宽敞明亮的开放空间。此外，这个展厅原本通过几段窄小的台阶通向面对金海湖的主展厅。我们把台阶放大，空间相互贯通变成一个整体。室内中心

are also arranged around it, implying that specific individuals is at the centre of city construction. The faded dark wood walls and ceilings are whitened, and the floor is redecorated with light terrazzo. We hope that Xianlu will continue to function as a public service facility by the lake after the site project. The landscape is renovated at the same time with the building. During the exhibition, what used to be a closed waterfront becomes an urban public space open to citizens. I believe that this is exactly the unique significance of SUSAS: recreate and redesign places through public events, and turn static exhibitions into dynamic spatial practices.

The opening forum of the site project "New City Regeneration", co-organized with Pin Zhu Architectural Forum of Shanghai Urban Planning and Natural Resources Bureau and facilitated by the journal *Urban China*, invites professionals and governmental authorities to discuss issues about new cities together. "New City Regeneration" involves two meanings. First, our understanding and definition of "new city" is continually updated. Second,

部位的会议室被用作纪录片的播放空间，规划与建筑设计的案例展品围绕它展开，也暗示了在城市建设过程中是具体的个人处于中心位置。墙面和吊顶已经陈旧的深木色被翻新成白色，地面重新做了浅色的水磨石。我们希望在展览之后，这个建筑能作为湖畔的公共服务设施继续使用。与建筑翻新同步，建筑所在场地也相应更新了景观。展览期间，这个原本封闭的滨水空间被打开，成为对市民开放的城市公共空间。我认为这恰巧是上海城市空间艺术季的独特意义，即以一种公共事件的方式重新创造和定义场所空间，把静态的展览转变为一种积极的空间实践活动。

此外，展览联合上海市规划和自然资源局的"品·筑建筑评论"组织了开幕论坛。论坛以"新城更新"为名，在《城市中国》杂志的协助下，邀请专家学者和政府职能部门共同讨论新城问题。"新城更新"有着两层含义：一方面意味着对"新城"的认识和定义的不断更新，另一方面还意味着高速发展和变化的时代，即便是新城，在很短的时间内也面临着城市更新的问题。论坛邀请了华东师范大学城市与区域科学学院的孙斌栋教授和深圳大学建筑与城市规划学院研究员张宇星博士作为主题发言人。孙斌栋教授是柏林工业大学城市与区域规划博士，他以《上海新城发展及其未来方向》为题，在宏观层面梳理了城市架构、开发体制、交通模式、人口和就业等新城开发战略问题。张宇星博士是深港城市／建筑双城双年展的主要发起人，他的报告主题为《新一代城市模型：品牌塑造对新城建设的持续效应》。张博士从"城市策展"与"策展城市"的趋势出发，强调了展览对于促进城市空间经济的独特作用。主题报告之后，论坛还邀请了同济大学建筑与城市规划学院李振宇教授、章明教授、田宝江教授，以及知名建筑师俞挺、张佳晶、陈嘉炜等嘉宾共同就奉贤新城的规划、建筑、城市法规等层面分享了各自的见解和建议。

in this era of rapid development and changing, even a "new" city will face its own regenerative issues in a short time. The keynote speakers are Prof. Sun Bindong of School of Urban and Regional Science, East China Normal University and Dr. Zhang Yuxing, researcher of School of Architecture and Urban Planning, Shenzhen University. Sun, an PhD of urban and regional planning in Technische Universität Berlin, offers a macro survey of city structure, developmental mechanism, transportation model, population, employment and other strategic issues of new city development in *Shanghai New City Development and Its Future Directions*. Zhang, the primary initiator of Bi-city Biennale of Urbanism / Architecture, gives a speech on *New City Model: Consistent Effects of Branding on New City Construction*, emphasizing the unique roles that exhibitions play in promoting urban economies based on the trends of "Curation in City" and "Curating City". After the keynote speeches, Profs. Li Zhenyu, Zhang Ming and Tian Baojiang of College of Architecture and Urban Planning Tongji University, and renowned architects Yu Ting, Zhang Jiajing and Chen Jiawei, and other honored guests share their insights and suggestions for the planning, building and regulation of Fengxian New City.

城市蜕变——虹口区实践案例展

City Metamorphosis — Site Project in Hongkou District

苏杭
香港大学上海学习中心公共项目主任、策展人

Su Hang
Director of Public Projects and Curator, Shanghai Study Centre,
The University of Hong Kong

主办单位：
上海市虹口区人民政府
承办单位：
虹口区规划和自然资源局、虹口区文化和
旅游局、虹口区北外滩街道
策展团队：
香港大学上海学习中心
展览时间：
2019 年 10 月 31 日—2020 年 1 月 6 日
展览地点：
虹口区北苏州路 298 号

Host:
People's Government of Hongkou District, Shanghai
Organizers:
Hongkou District Planning and Natural Resource Bureau,
Hongkou District Culture and Tourism Bureau, Hongkou District
North Bund Sub-district Office
Curatorial Team:
Shanghai Study Centre, University of Hong Kong
Time:
31 October 2019 – 6 January 2020
Venue:
298 North Suzhou Road, Hongkou District

2019 上海城市空间艺术季虹口区实践案例展与往届不同。今年展览分为三个大板块，即"剩余空间：苏州河口地区滨水空间教学及研究""生活提案""城市蜕变"，以苏州河北岸的香港大学上海学习中心联动北外滩白玉兰广场、建投书局（浦江店）、2019 空间艺术季主展场的小白楼报告厅举办的一系列公众活动，串联起苏州河—黄浦江的滨水空间，邀请公众参与"我的城市徒

The SUSAS 2019 site project in Hongkou District, different from its previous counterparts, consists of three panels, "Residual Space: Teaching and Researching of Waterfronts at the Mouth of Suzhou Creek", "Urban Life Generator", and "City Metamorphosis". A set of waterfront spaces along Suzhou Creek and Huangpu River are connected by a series of public activities held in Shanghai Study Centre, University of Hong Kong on the northern bank of Suzhou Creek, Shanghai Magnolia Plaza at North Bund, JIC Bookstore Pujiang and Little White Building Lecture Hall as part of the SUSAS 2019 main exhibition, inviting citizens to take part in "Stroll in My City Project", get involved in

31/10
王林
薛鸣华
【开幕讲座】河畔客厅、共享水库 - 苏州河滨水设计
地点：香港大学上海学习中心

23/11
高亦陶
【沙龙】城市蜕变 - 剩余空间与公共性
地点：上海城市空间艺术季主展场小白楼报告厅

17/11
无微不至——
社区微更新的
多维度解读
【沙龙】
地点：香港大学上海学习中心

26/11
Minsuk
CHO
【英文讲座】Challenges
地点：香港大学上海学习中心

08/12
戴邹君
【沙龙+城市徜步】觅境·北外滩历史建筑鲜为人知的故事
地点：白玉兰广场金领驿站

22/11
朱晓明
【讲座】力之力——上海滨水工业遗产的感知与记忆
地点：上海城市空间艺术季主展场小白楼报告厅

14/12
品读历史建筑，延续城市文脉：
历史人文建筑活化与城市规划
赋能
【沙龙+闭幕论坛】
地点：建投书局·浦江店

City Metamorphosis: Renewal of Waterfront Space, Landscape and Community
城市蜕变：滨水空间，景观与社区的更新
2019 上海城市空间艺术季虹口区实践案例展 / Shanghai Urban Space Art Season Site Exhibition of Hongkou District / 10 月 31 日–1 月 6 日 / October 31-January 6 / 香港大学上海学习中心展厅艺术廊 / Ground Floor Gallery of Shanghai Study Center (HKU) / 虹口区北苏州路 298 号 / 298 North Suzhou Road, Hongkou District

虹口区实践案例展展期活动海报
Poster of Hongkou Site Project

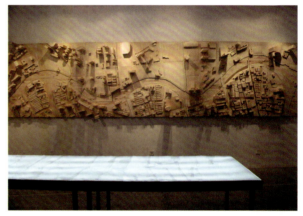

"剩余空间"展场
Venue of "Residual Space"

步计划"，参与并见证"一江一河"城市蜕变策略。

剩余空间：苏州河口地区滨水空间教学及研究

主展览梳理了香港大学上海学习中心多年来对北外滩及虹口区其他地方的研究与教学成果，从城市形态、历史、文化比较、资本业态、居民个体内心与意识、社会问题等多个主题，以建筑学的独特视角为观者打开进入历史、文化、艺术和当下现实空间语境的渠道。展览将作为苏州河畔的公共艺术项目介入一江一河沿线的城市发展之中，在呈现城市、社区、景观三个层级的研究和项目的同时，邀请公众共同探讨亚洲滨水城市发展的可能性。香港在城市发展和更新的过程中，

and witness the city metamorphosis strategies of Huangpu River and Suzhou Creek.

Residual Space: Teaching and Researching of Waterfronts at the Mouth of Suzhou Creek

The main exhibition demonstrates the educational and academic efforts devoted to the Bund and elsewhere in Hongkou District by Shanghai Study Centre. Urban form, history, cultural comparative study, business form, consciousness of individual residents, social issues will be discussed from the perspective of built environment, channeling history, culture, art and current spatial contexts. The exhibition will be integrated into the development of waterfront Shanghai as a public art project by Suzhou Creek. The exhibited projects and studies on three levels – city, community, landscape – will invite citizens to explore the developmental possibilities of Asian waterfronts. As Hong Kong develops and renews itself, fragmented public spaces and

碎片化的公共空间与资源之间最终会达到动态的平衡。香港大学上海学习中心将"剩余空间"的研究方法应用于设计教学与城市研究中，并持续观察上海中心城区，尤其是滨水地带在城市、景观与社区空间中独特的发展现象。

基地研究模型

在城市发展和更新的过程中，由于经济、管理、历史等原因，有限的空间资源常常难以获得最科学的分配方式，这便导致城市中出现许多"剩余空间"。香港大学建筑学院夏校项目的学生通过制作基地模型，对苏州河两岸的物理空间展开记录。基地范围西起恒丰路，东至黄浦江河口，学生使用模型分析出不同尺度和层次的剩余空间问题，并对这些城市问题进行由微观至宏观的探究。

resources will always reach a dynamic equilibrium. Shanghai Study Centre, The University of Hongkong decides to teach design and study cities using the idea of "Residual Space" while observing on a continual basis how the spaces, landscapes and communities in central Shanghai, especially waterfronts, have developed.

Site Model

Due to economic, administrative, historical and other factors, a developing and regenerating city hardly assigns its limited spaces in most scientific ways, and that's how so many "residual spaces" appear. The students enrolled in the summer school of Hong Kong University Faculty of Architecture have, by producing a site model, recorded the physical spaces on both banks of Suzhou Creek. The site starts from Hengfeng Road and extends eastward until the mouth of Huangpu River. The students use the model to analyze the issues of residual spaces on multiple scales and levels, and explore these issues in both micro and macro perspectives.

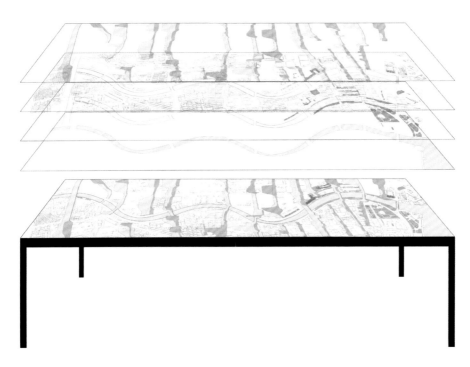

城市"透"视
Perspectives of city

观众参观苏州河口木质基地模型
A visitor watching the Wooden Site Model of the Mouth of Suzhou Creek

城市"透"视

由于"剩余空间"在各方面均没有界定，研究者需要通过设立参考系对剩余空间进行认知和研究。这张分析图实则由四层图纸叠合而成，以尺度、归属权、利用情况不同的城市界面作为参考系，与城市的物理空间、街区肌理、景观要素叠合，研究剩余空间与各要素之间的关系。参观者可以通过这张分析图，对墙上抽象的模型进行具象的阅读。该分析图也为后续进一步的研究和设计提供可能性。

"蜕变"的历史——苏州河畔

在历史上，由于长期被定义为上海行政区划的界河，苏州河逐渐成为了行政管理上的"剩余空间"，这一特性直接导致苏州河两岸南北发展上的明显差异，以及沿岸公共空间的逐渐衰落。展览通过历史地图，展示了苏州河由 1504 年至 1949 年数百年间的蜕变历程，分析了苏州河在城市发展演进过程中，如何持续扮演着边界的角色。

生活提案：滨水空间节点的解读和生活的未来

展览的第二个板块是针对苏州河口滨水空间节点的解读、记录和设计畅想。在本次展览中，我们选取了六个节点，从历史文脉、路

Perspectives of City

Since the range of "residual space" is generally undefined, researchers need to set up references to understand and study it. This analytical diagram is an overlay of four drawings. Given the references of scale, entitlement, usage and other urban interfaces, and superimposing factors of physical space, street fabric and landscape, it is able to study the relationship between residual spaces and other factors. The diagram allows the visitor to read the abstract model on the wall in a concrete manner, and makes possible further researches and designs.

A History of Metamorphosis — Banks of Suzhou Creek

Since Suzhou Creek is for a long time a boundary between administrative districts, it gradually becomes an administrative "residual space", directly leading to the obvious differences between its northern and southern banks in terms of urban construction, and the increasing decline of its waterfront public spaces. At the site project, the centuries-long metamorphosis of Suzhou Creek between 1504 and 1949 are demonstrated with historical maps, and the issue of how Suzhou Creek, with the evolution of Shanghai, may continue to play its role of boundary is analyzed.

Urban Life Generator: Interpreting Waterfront Nodes and Future of Life

The second panel focuses on interpreting, recording and imagining the waterfront nodes at the mouth of Suzhou Creek. Six selected nodes are used to observe and analyze the current features from the perspectives of history, road network, urban interface, infrastructure and waterscape, and conceptual vitalizing designs are proposed accordingly.

"生活提案"展场
Venue of Urban Life Generator

网系统、城市界面、基础设施和亲水景观的角度来观察和分析现有空间的特征,并以此为基础,提出概念性的激活提案。

城市界面与河滨休憩平台

通过对历史的研究,研究者发现由于苏州河长期被定义为行政区划的界河,直接导致了苏州河南北两岸发展上的巨大差异。两岸最大的差异,则是围合公共空间的城市界面,在轴向分布上的明显不同。苏州河南岸,沿着圆明园路、虎丘路的历史建筑以及城市界面均沿南北纵向排布,而北岸则恰恰相反,沿东西横向排布。设计从该差异中提取出概念和要素,在场地中放置不同轴向的"界面",通过对河滨空间的连接和阻隔,重新协调和激活苏州河沿岸公共空间。

城市剧院

研究者发现,在苏州河沿岸的街区中,剩余空间往往存在于道路系统的交汇处。人行道、非机动车道、机动车道、河道等不同级别的道路系统的交汇处,剩余空间的尺度、归属等各方面也有所不同,但总体上作为城市中的消极空间,被人们所忽视。设计引入"露

Urban Interface and Waterfront Leisure Deck

Historical researches show that the direct cause of the differences between the northern and southern banks of Suzhou Creek, is the fact that Suzhou Creek has long been defined as an administrative boundary. The biggest difference is that the urban interfaces enclosing waterfront public spaces are distributed along two distinct axes. On the southern bank, the historical buildings and urban interfaces along Yuanmingyuan Road and Huqiu Road run in a vertical, or south-north direction, while the condition of northern bank is exactly the opposite. Thus the design proposal, retrieving concepts and elements from this difference, places "interfaces" along various axes, and tries to coordinate and activate again the public spaces along Suzhou Creek by connecting or blocking certain sections of the waterfront.

City Cinema

Researchers find that in the blocks along Suzhou Creek, residual spaces tend to be at intersections. While the scale and entitlement of these residuals vary with the type of intersection – sidewalk, non-vehicle road, vehicle road or river course – they are in generally ignored as passive urban spaces. The new function of "open cinema" introduced in the design is an intervention into urban space management and generation, activating the "blind spots" in an acupuncture-like manner.

City Atrium

Researchers find that the urban fabric along Suzhou Creek is

天影院"作为新的功能，介入城市的空间管理和生产，对这些城市"盲点"进行针灸式的激活。

城市中庭

研究者发现，苏州河沿岸的城市肌理存在"多孔性"的特征，而公共空间在不同尺度上，也与周边的环境和建筑存在着相互"渗透"的关系。在苏州河南岸的历史街区和里弄中，存在着大量尺度小但连续性强的公共空间，相比之下的北岸，大体量的建筑中，往往存在尺度大、但较为分散和碎片化的公共空间，如邮政博物馆内部的巨大中庭和河滨大楼中心的公共庭院。设计提出"城市中庭"的概念，沿苏州河岸布置连廊和休憩亭，暗示和引导使用者去发现和使用隐藏在巷弄和建筑中的"剩余空间"，并和河滨走廊重建连续性。

城中之隙

研究中发现，苏州河两岸有大量的"剩余空间"存在于建筑间的"缝隙"之中。这些"剩余空间"原本只是城市的"边角料"，是在城市规划和建设过程中产生的未使用或者使用权不明的小地块，但在城市发展过程中，已被人们以各种方式错用、滥用和占用了。通过分析，我们总结了三种自发占用"剩余空间"的模式：底层功能性加建（如商铺、储藏室），中层连接性构筑物（如连桥）以及顶层遮盖物（如悬挑的露台）。设计在这些自发的占用模式的基础上，发展出新的原型，使得这些存在于"城中之隙"的公共空间可以更好地被利用起来。

水上环形休憩步道

在城市蜕变的过程中，苏州河已基本丧失原

"porous", and that the public spaces in various scales "infiltrate" their adjacent environment and buildings, and vice versa. In contrast to the southern bank whose historical blocks and lanes are rich in small yet continuous public spaces, the public spaces among the massive buildings on the northern bank — for example, the gigantic atrium inside Shanghai Postal Museum and the central courtyard of Riverside Building — are usually large yet scattered. The proposed concept of "city atrium" consists of continuous terraces and rest pavilions along Suzhou Creek, hinting and guiding users to find and use the "residual spaces" hidden among lanes and buildings, and thus reestablishes a continuous waterfront corridor.

Gaps in City

Researchers find many "residual spaces" among the "gaps" between buildings on both banks of Suzhou Creek. These spaces, original the small plots that are unused or unentitled in the process of urban planning and construction, have already been misused, overused or occupied in various ways as the city develops. Based on our analysis, we believe that there are three modes of spontaneous occupation of "residual spaces": ground-floor utilization (shop or storage), middle-floor connection (cat walk) and top-floor covering (long-hanging balcony). In our design proposal, new prototypes are developed based on these spontaneous modes so that these "gaps" can be better utilized.

Strolling Circle

With the metamorphosis of Shanghai, Suzhou Creek has largely lost its original function as a canal and hence become the largest "residual space" in the city. There were many factories and shops clustered along Suzhou Creek by 1960, and as an essential transporting hub, the river used to be a key element of public spaces and daily life. The design proposal suggests to build a circle of strolling lanes above Suzhou Creek, reconnecting the river with the life of residents and forming a "chain-circle" of waterfront public landscapes as an approach to the general revitalization of public spaces along Suzhou Creek.

City metamorphosis: Suzhou Creek, Huangpu River and Individual Participation

观众体验我的城市徒步计划，打卡盖章创作台
The creative marking desk of "Stroll in My City Project"

戴邹君博士导览北外滩历史建筑徒步活动
Tour guide of historical buildings in North Bund led by Dr. Dai Zoujun

虹口区实践案例展闭幕论坛
Closing Forum of the Hongkou Site Project

本作为运河的功能，成为城市中最大的"剩余空间"。1960 年前，大量的工业以及商业仍聚集在苏州河两岸，作为重要交通枢纽的苏州河曾是公共空间和生活的重要环节。设计提出在苏州河上，建立一个水上环形休憩步道，重建居民生活与苏州河之间的关系，形成"环链状"沿河公共景观区域，以此实现苏州河沿岸公共空间的整体性复兴。

During the period of site project, seven public events are held where more than ten domestic or international researchers and practitioners in the fields of planning, architecture, landscape, art and culture are invited to share their studies and practical experiences about waterfront spaces with local residents. The exhibition hall has a marking point where all visitors can leave their ideas and feelings about the exhibition. This creative desk is welcomed by citizens of all ages: there are "stampers" in their seventies who come here specifically for the stamps of historical buildings, and teenagers aiming to compose their sketchbooks. "Stroll in My City Project" and "Tour of Historical Buildings in

城市蜕变：一江一河与个体参与

在案例展展期过程中，我们安排了七场展期公共活动，邀请十余位海内外规划、建筑、景观、艺术、文化等方面的学者和专家，同居民分享他们对滨水空间的研究和实践经验。展厅内还有专门为参观者设立的打卡角，在这里所有参观者都可以自由创作，留下对这次展览的感受和心情。创作台受到了不同年龄段的市民的欢迎和喜爱，有七十多岁的"敲章族"专门来收集历史建筑印章，也有十几岁的高中生带着自己的手账前来创作。"我的城市徒步计划"与展期活动"觅境·北外滩历史建筑漫步"相结合，希望社区居民、公众以及对虹口、北外滩有兴趣的各界学者，将这个小小的创作台作为一个起点，去探索上海滨水空间中散落的建筑故事。

展后感想

2019年的上海城市空间艺术季（以下简称SUSAS）和前两届相比，规模、观展人数、展览本身的内容深度和涉及话题的广度都达到了更高的层次。2017年的时候，SUSAS更像是一场面向市民的期末汇报，从主展到各区案例展，大多是对过去两年内城市更新、区域规划的项目、研究做一次集中展示。今年的SUSAS做了一些"跨界"的尝试，以规划系统为主，联动多级政府部门、民间组织、高校、社区来集中探讨我们的城市正在面临的问题和可能的解决方案。对于居住在城市中的人们来说，公共资源的价值不仅仅是某个空间的品质提升，还包括对都市生活共同愿景的更新。SUSAS从某种程度上来说，像是一剂催化剂，或者是一个推手，把整个城市发展的愿景呈现在公众的面前，让不同背景、不同年龄的人可以理解城市更新的意义，其中会有一些人加入这个进程。SUSAS是由政府层面牵头、组织、指导的艺术季，同时也在为未来更多民间的、自发的更新力量提供着启发和帮助。

放眼全球，重要城市的建筑、设计艺术季有二三十个，但更多的是以一种专业内部的话语来讨论建筑、城市的未来是什么。2017年我对SUSAS的评论里曾提到，尽管策展

虹口区规划和自然资源局罗翌弘副局长致辞
Address of Luo Yihong, deputy director of Urban Planning and Natural Resources Bureau of Hongkou District

虹口区实践案例展闭幕论坛观众提问
FAQ session of the closing forum of the Hongkou Site Project

North Bund" together reflect the hope that this modest creative space may serve as a starting point from which local residents, the public and researchers interested in Hongkou and North Bund would explore the architectural stories scattered among Shanghai waterfronts.

Reflections

Compared to SUSAS 2015 and 2017, SUSAS 2019 reaches a higher stage in terms of size, number of visitors, depth of exhibited contents and range of topics involved. Back in 2017, SUSAS was more like a final presentation to citizens – its main exhibition and site projects were largely a concentrated demonstration of recent projects and researches about urban regeneration and regional planning in the previous couple of years. This year, SUSAS has made some "hybrid" attempts where governmental agencies of different levels and professions, NGOs, universities and communities, with the planning authorities as the mainstay, are mobilized to discuss the problems our cities are facing now and their potential solutions. For city dwellers, public resources are valuable not only in terms of improvement of this or that space, but a renewed common vision of urban life. In a sense, SUSAS is a catalyst or a motivator that presents an overall vision of the entire city before the public and allows people of various backgrounds or ages to understand the significance of urban regeneration, some of whom might even partake in the process. As an event led,

方引进了很多同期国际展览的优秀作品，但并未将学科的宽度进一步打开。如果策展理念从建筑学出发，配合社会学、人类学、哲学等领域空间转向的发展趋势，是否可以形成跨学科的多层次讨论？

以2019SUSAS虹口区实践案例展"城市蜕变"来说，虹口区规划和自然资源管理局、虹口区文化和旅游局联合北外滩街道共同举办，并邀请了香港大学、上海交通大学、同济大学、上海大学及上海美术学院等高校教授专家，为广大市民分享他们的设计、研究以及对城市的思考。这场历时不足3个月的展览，配合7场市民活动贯通"苏州河—黄浦江"北岸，将城市本身所承载的历史文化意义和滨江地带得天独厚的区位条件融合在一起，为公众带来时间和空间交融的立体观展体验。

从规划依赖到艺术兴城，SUSAS用五年的时间和一整座城市来讲述城市更新的故事。我们站在当下回望2019，又生出些许新的思考和担忧。无论是日常的整体管控，还是突发的危机应对，对已经进入"公共愿景式更新"的上海来说，已经交上了一份不错的答卷。未来如何利用SUSAS的推动作用来应对城市更新工作？如何使文化事件所带来的公众关注得以形成持续发酵的创新动力？如何更有效地利用散落社区、街道的微空间来孵化民间的研究组织，为展览持续注入多样性？希望未来有更多公众带着他们的学科智慧加入对城市问题的发现与思考之中，他们将是这些问题最具智慧的回答者。

organized and guided by the public sector, SUSAS also inspires and supports the spontaneous, civil forces of regeneration.

There are about twenty or thirty architectural or design art seasons in major cities around the world, yet the larger part of them are discussions about the future of buildings and cities within professional circles. As I commented about SUSAS 2017, while the curators introduced many excellent works of contemporary international exhibitions, they failed to conceive a broad interdisciplinary vision: is it possible to establish a multi-level interdisciplinary discussion by starting from an architectural perspective in curation and turning to sociological, anthropological and philosophical trends about spaces?

The SUSAS 2019 site project in Hongkou District "City Metamorphosis" is such an example. It is co-hosted by Urban Planning and Natural Resources Bureau of Hongkou District, Culture and Tourism Bureau of Hongkou District and Beiwaitan Sub-district, and professors of Hong Kong University, Shanghai Jiaotong University, Tongji University, Shanghai University with Shanghai Academy of Fine Arts are invited to share their designs, researches and reflections on cities. This site project lasting for less than three months connects, in collaboration with 7 public activities, the northern bank of Suzhou Creek and Huangpu River. It integrates the inherent historical and cultural significance of the city with the uniquely prominent conditions of waterfront, and provides the visitors with an enriched experience in which times and spaces are entangled.

In its history of five years and based on the entirety of Shanghai, SUSAS has been telling stories of urban regeneration from planning to public art. Now it's 2020, and in retrospect of 2019, we have some new worries and concerns. Shanghai, in its stage of "renewal of common vision", has fared pretty well in both daily general administration and emergency handling. In the future, how SUSAS may continually promote urban regeneration in Shanghai? How can turn the public attention aroused by cultural events into a consistent motivation for creativity? How micro-spaces scattered among communities and blocks may serve as effective incubators of non-governmental research organizations that keep adding diversity to exhibitions? We hope that a wider public will contribute insights from their own disciplines to looking for and thinking about issues` facing our cities, and we believe that they are the wisest answerers to these questions.

水之魔力——闵行区浦江第一湾公园
实践案例展

Magic of Water — Site Project at Pujiang First Bay Park, Minhang District

张海翱
上海交通大学设计学院副教授
奥默默工作室主持设计师

Zhang Hai'ao
Associate Professor of School of Design, Shanghai Jiaotong University, Chief Designer of 'Oumoumou Studio

主办单位：
上海紫竹高新技术产业开发区、上海市闵行区吴泾镇人民政府
承办单位：
上海交通大学设计学院、Let's Talk 学术论坛
协办单位：
闵行区规划和自然资源局
策展人：
张海翱、戴春
展览时间：
2019 年 11 月 24 日—2020 年 1 月
展览地点：
浦江第一湾公园

Hosts:
Zizhu National High-tech Industrial Development Park, People's Government of Wujing Town, Minhang District, Shanghai
Organizers:
School of Design, Shanghai Jiaotong University, Let's Talk Forum
Co-organizer:
Urban Planning and Natural Resources Bureau of Minhang District
Curators:
Zhang Hai'ao, Dai Chun
Time:
24 November 2019 – January 2020
Venue:
Pujiang First Bay Park

"浦江第一湾"是"浦江合流"工程的起点，其终点是今日的陆家嘴。上海市闵行区民间文艺家协会主席张乃清先生认为，从"第一湾"到"陆家嘴"，可以领略整个上海千年历史的发展，"第一湾"是上海千年历史的起点！华东师范大学王家范教授认为，上海是以其腹地经济的雄厚和水运四通八达的"地利"，成就其作为全国经济中心之大业。紫竹高新区作为上海市黄浦江沿岸唯一的国家高新区，毗邻近七公里滨江岸线，恰好位于"浦江合流"工程的起点，"浦江第一湾"以其强大的历史、文化、地理、生态特色赋予了紫竹地区独特的文化认知和标杆意义。二者分别作为闵行滨江的生态符号与科技符号，进行同期建设与发展，引领时代发展趋势；同时，上海罕见约 40 万平方米的大型人工湖、闵行第一湖——兰香湖，也已建设完工，将黄浦江腹地向内延展了一公里，力图最大化浦江第一湾资源。未来，"浦江第

Pujiang First Bay is the starting point of "Convergence Project of Huangpu and Wusong River" which ends at what is now known as Lujiazui. Zhang Naiqing, president of Minhang Folk Literature and Art Association, believes that the thousand years' history of Shanghai is embodied between Pujiang First and Lujiazui, and Pujiang First Bay is where the millennial history of Shanghai began! According to Prof. Wang Jiafan of East China Normal University, what makes Shanghai the national economic hub is its prosperous hinterlands and convenient water transportation. Zizhu National High-tech Industrial Development Park, as the only national high-tech development zones by Huangpu River with a waterfront of almost 7 kilometers, happens to sit at the starting point of "Convergence Project of Huangpu and Wusong River" and thus benefits from the distinctive historical, cultural, geographical and ecological characteristics of Pujiang First Bay, making itself a unique cultural landmark. The Development Park as the technological symbol, and Pujiang First Bay as the cultural emblem of Minhang District are simultaneously constructed and developed, leading the tides of time. And the largest lake in Minhang, Lanxiang Lake (400,000m²) which is also an unusually immense artificial lake in Shanghai, is completed at the same time. It extends Huangpu River towards its hinterlands

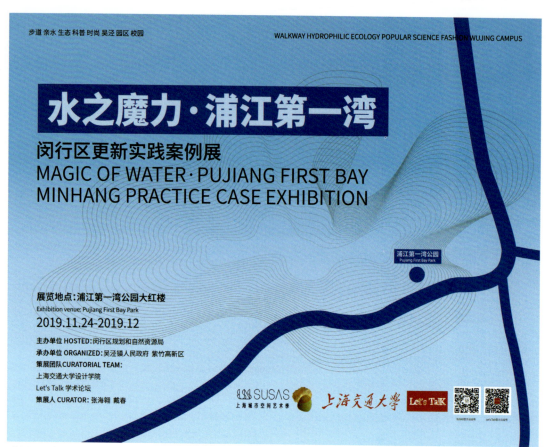

"水之魔力"展览海报
Poster of "Magic of Water"

一湾公园"规划建设，将在规划导向、建设内容方面进行科学设置、有序引导，发挥紫竹国家级高新区应有的能级与影响力，塑造上海市的世界级城市历史文化"新"地标。

随着闵行版外滩的蓝图逐步走向现实，"浦江第一湾"滨水空间逐渐走向集人文、科技、生态价值于一体的高新区游览体验空间。

主题阐述

黄浦江是上海最大的河流，也是长江入海前的最后一条支流。黄浦江干流的起点在淀山湖湖口，由西向东进入上海市，在上海市的中部转向，由南至北奔向东海，这一呈90°的直角大转弯，就是"浦江第一湾"。
"水之魔力——闵行区浦江第一湾公园实践案例展"充分利用浦江第一湾公园"步道、亲水、生态、科普、时尚、吴泾、园区、校园"八大优势，全方位彰显闵行的活力，展现其中的魅力，将水的特点、变化、作用等

for about 1km as an attempt to maximize the available resources of Pujiang First Bay. In the future, the planning and construction of Pujiang First Bay will be scientifically arranged and orderly guided to fulfill the energies and influences which Zizhu National High-tech Industrial Development Park should have, and to create a "new" world-class historical and cultural landmark in Shanghai.

With the gradual realization of the Minhang version of Bund blueprint, Pujiang First Bay is becoming an experiential destination based on a high-tech development zone which combines cultural, tech and ecological values.

Theme Interpretation

Huangpu River is the largest river in Shanghai and the last tributary of Yangtze River before it meets the sea. The trunk stream of Huangpu River originates from the mouth of Dianshan Lake and flows eastward into Shanghai before it turns about in the middle of the city and runs northward into East Sea. The square turn is called Pujiang First Bay where Pujiang is shorthand for Huangpu Jiang, or Huangpu River. "Magic of Water — Site Project at Pujiang First Bay Park, Minhang District" takes fully advantage of the eight assets of Pujiang First Bay

滨水主题论坛
Waterfront Forum

完美结合。

策展方案

闵行区吴泾镇所处的"浦江第一湾"历史上是"江浦合流"工程的起点。该工程直接促成了黄浦江的形成，改变了上海的地理格局，为上海千年来的城市发展奠定了基础。如今吴泾通过不断的环境改造与产业升级，以科技和时尚作为主导产业，依托紫竹高新区，以及上海交通大学、华东师范大学，还有镇域内各大产业园区的科技研发能力，走出一条面向未来的智慧社区和智能城区的新路。

闵行区浦江第一湾公园实践案例展及吴泾历史回顾展开幕

11月24日，2019上海城市空间艺术季系列案例展之"水之魔力·浦江第一湾——闵行区更新实践案例展"在第一湾公园大红楼开幕。

在"浦江第一湾"的策展开幕式中，举行了图片展、高峰论坛、参观导览和讲座等一系列活动。代表人简单致辞后，开幕现场举行了"泾"彩四溢——当代走秀艺术舞蹈，飞行家身体剧场的艺术家们以亲水平台为舞

Park — promenade, waterfront, ecology, science popularization, fashion, Wujing River, high-tech development park and university campus — and fully demonstrates the vitality and charm of Minhang while incorporating the features, variations and effects of water.

Curatorial Program

Pujiang First Bay at Wujing Town, Minhang District is the starting point of the ancient project of converging Huangpu and Wusong River, which directly led to the forming of Huangpu as a major river and changed the geography of Shanghai, laying the foundation for the development of Shanghai in a thousand years. Wujing Town has been continuously transforming its environment and upgrading its industries. Based on Zizhu National High-tech Development Park and the R&D capacities of Shanghai Jiaotong University, East China Normal Universities and its other industrial parks, Wujing is walking on a new path featuring smart community and urbanity.

Opening Ceremony of the Site Project at Pujiang First Bay Park, Minhang District and Exhibition of the History of Wujing Town

On 24 November, the SUSAS 2019 site project "Magic of Water — Pujiang First Bay, Minhang Practice Case Exhibition" was opened at the Red Building in the park.

The opening ceremony consists of an image exhibition, a summit, a guided tour and lectures. After the short address of the representative is "Shining Wujing – Modern Catwalk Artistic

随着楼梯行进不断展开的吴泾历史图片展
Exhibition of the History of Wujing Town that unwinds with the staircase

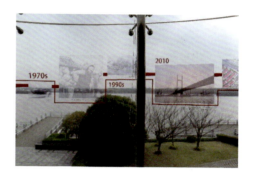

吴泾历史图片展
Exhibition of the History of Wujing Town

台，使用歌舞应和的方式演绎沉浸式现代舞作品《溯源》，带领观众完成了一次难忘的滨江艺术导览。

在接下来的滨水主题论坛中，专家们讨论了滨水工业空间转型与科技产业园区建设的主题。来自新加坡盛裕、潘冀联合建筑事务所及水石设计的专家分享了新加坡、苏州工业园区、新竹科学园区及更多科技主题园区的实践；来自中建工程、紫竹高新区、英国Spark 及闵行区管理部门的专家共同讨论高新技术产业园区的发展方向、建设路径与建立相关各常效沟通机制的可能方式。

Dance" where the artists of Feixingjia Dance Theater perform the immersive modern dance *Tracing the Roots* on a waterfront platform, leading the audience through a memorable art tour of Huangpu River amidst songs and dances.

The next activity is a forum in which professionals discuss the transformation of waterfront industrial areas and the construction of high-tech industrial parks. Experts from Surbana Jurong (Singapore), JJP Architects and Planners, and Shuishi share their experiences in Singapore, Suzhou Industrial Park, Hsinchu Science Park and other tech parks; and experts from China State Construction Engineering Corporation, Zizhu National High-tech Industrial Development Park, Spark (UK) and Minhang administration authorities discuss the directions and approaches of high-tech industrial parks, as well as possible ways to

"水之魔力"展览期间活动
Activities during the site project "Magic of Water"

东丽绿色创新—水处理膜—反渗透膜元件的结构（图片来源：网络）
Green Innovation – water treatment – structure of RO elements (source: Internet)

9:30—13:00，围绕"浦江第一湾"公园开展了"无奔跑不青春"滨江慢行彩虹跑活动。各局领导共同参加，进行了绕公园 10 公里不分男女老少的奔跑活动。

接下来的 3 天时间以吴泾历史回顾展为主。红楼楼梯行进不断展开，象征着每个时代广大人民群众"拾级而上"共同努力，吴泾的发展史娓娓道来。用一根红线串联起悠久的历史，与树林中的红楼，远处繁忙的河道融为一体。

在此之后的每一周活动，展览方邀请了来自覆盖设计、艺术、政策、地产、传媒的 5 位全领域专家嘉宾，包括学校建筑城规领域专业教授、艺术领域大咖、媒体和传媒大咖等作为客座讲师。活动包括交大景观植物工作营、交大历史工作营、交大儿童绘本活动、浦江第一湾鸟类观察工作营、可食水系植物科普工作坊、观察水位变化记录工作坊、可口可乐展览、东丽讲座、滨江徒步抖音VLOG 摄影活动、亲子森林 T 台秀、创意服装工作坊等。

establish permanent and effective communicative mechanisms among stakeholders.

9:30-13:00: the waterfront jogging activity "Run Youth" around Pujiang First Bay Park (10km) is attended by diversified participants, including leaders of multiple bureaus.

The following three days focus on Exhibition of the History of Wujing Town. The progressive staircase symbolizes the combined efforts of the people in various times to "climb the steps", and the stories of Wujing are told along the way. A long history is thus connected by a red thread, integrated with the red building among the trees and the busy river course far beyond.

From then on, every week the host offers activities with invited professionals in the fields of design, art, public policy, real estate and media, including guest lecturers who are architectural or urban planning professors, established artists and renowned figures in media circles. These activities include: botanic workshop supported by Shanghai Jiaotong University; historical workshop supported by Shanghai Jiaotong University; picture book workshop supported by Shanghai Jiaotong University; bird observation workshop at Pujiang First Bay; workshop on edible water plants; water level recording workshop; Coca-Cola exhibition; lecture on Toray; TikTok photography competition; parent-child T-stage show in forest; and, among others, creative clothes workshop.

参展艺术家合照
Participating artists

研学绿色环保循环水处理（东丽水处理研究所展示厅）

此次活动将由东丽水处理研究所水处理技术人员带领参观学习上海水处理过程及最新处理技术，并现场交流。东丽水处理研究所主要从事各种水处理膜技术的研究开发，其中包括海水淡化技术，生活污水、工业废水的循环再利用技术，饮用水深度处理技术等应用开发，同时协助中国地区水处理膜产品的技术支持，以此推动先进的膜技术在中国市场的应用，并与国内一流科研机构开展共同研究，积极参加国家和地区的重点科研项目。

体验高新技术产业发展（上海紫竹国家高新技术产业开发区）

上海紫竹国家高新技术产业开发区由闵行区人民政府、上海交通大学、紫江集团、上海联和投资公司等七家股东单位共同投资组建，于 2002 年 6 月 25 日奠基，一期规划面积 13 平方公里。2011 年 6 月，获得国务院《关于同意上海紫竹高新技术产业园区升级为国家高新技术产业开发区的批复》。自 2004 年始，高新区连续四届八年获得"上

Field Trip of Environmental Water Recycling Technologies (Exhibition Hall of Toray Research Institute of Water Treatment)

In this field trip, technicians of Toray Research Institute of Water Treatment lead visitors to watch water treatment processes in Shanghai and the latest treatment technologies, and discuss with the visitors there. Toray Research Institute of Water Treatment focuses on developing water treatment membranes, including seawater desalination, recycling of domestic sewage and industrial wastewater, and advanced treatment of drinking water. To promote the adoption of advanced membranes in China, the institute provides technical supports to water treatment membranes in this market. It also engages in collaborative programs with first-class Chinese research institutes and actively participate in regional or national key research projects.

Experience the Development of High-tech Industry (Zizhu National High-tech Industrial Development Park)

Zizhu National High-tech Industrial Development Park, funded by seven stockholders including People's Government of Minhang District, Shanghai Jiaotong University, Zijiang Holdings and Shanghai Alliance Investment Ltd., laid its foundation at 25 June 2002. The first-phase project is planned to cover an area of 13km^2. The State Council of the People's Republic of China, in June 2011, issued *The Approval of Upgrading Zizhu High-tech Industrial Development Park to a National High-tech Industrial Development Park*. In the eight years after 2004, it was

海市文明单位"称号。

展后感想、相关反响及启发

随着江河航运功能的相对弱化，城市的滨水工业空间逐步转型，面向新的产业与生活方式需求下的空间资源重组，一些滨水空间成为新兴科技产业园区的重要选择之一。无论是从空间资源重新分配视角，滨水空间利用的价值导向，还是城市总体格局发展的需求角度来思考这样的转型，都让我们看到城市转型阶段对于空间品质提升，空间合理化利用，以及开发模式的理性化发展的重视。

紫竹高新区对"浦江第一湾公园"的建设还在不断考虑和探索。这是一个历史悠久的区域，应该展示黄浦江第一湾的前世今生，让一代代的居民知道这里的故事，传递人文记忆。这也是一个欣欣向荣的区域，国家级的高新区完全有能力充分发挥在高科技、教育、文化、体育方面的优势，让更多的人，特别是孩子们体验历史第一湾所焕发出来的新时代独特魅力——这是一种碰撞与交织！

（本书未标明来源的图片，均由作者提供）

consecutively awarded four times with "Model Unit of Shanghai".

Reflection, Reaction and Inspiration

With the weakening of river transportation, transformation of urban waterfront industrial areas, and reorganization of available spaces given demands for new industries and lifestyles, some waterfronts have become important potential sites for new technological and industrial parks. No matter from which perspective – reallocation of spatial resources, guidance for waterfront utilization, or general development of city – it is apparent that improving urban spaces, reasonable utilization of spaces and rational evolution of development mode are emphasized in transforming cities.

The construction of Pujiang First Bay Park in Zizhu National High-tech Industrial Development Park is still a topic of reflection and exploration. The long history of First Bay should be demonstrated so that its stories and memories may be carried on through generations. This is also a prosperous area. The national development park is fully capable of realizing its advantages in technology, education, culture and sports, allowing more people, especially children to experience the unique charms of Pujiang First Bay in the new era – there is clash, but also interweaving!

(All the pictures not indicated credit information in this book are provided by the authors)

图书在版编目（ＣＩＰ）数据

相遇 . 2, 2019 上海城市空间艺术季案例展 /《相遇》
编委会编 . -- 上海：东华大学出版社 , 2021
　ISBN 978-7-5669-1962-5

Ⅰ . ①相… Ⅱ . ①相… Ⅲ . ①城市空间 – 公共空间 –
建筑设计 – 作品集 – 中国 – 现代 Ⅳ . ① TU984.11

中国版本图书馆 CIP 数据核字 (2021) 第 175001 号

相遇
2019 上海城市空间艺术季案例展

《相遇》编委会 编

策　　划：上海城市公共空间设计促进中心
　　　　　群岛 ARCHIPELAGO
特约编辑：辛梦瑶
责任编辑：高路路
设计排版：李高

版　　次：2021 年 1 月第 1 版
印　　次：2021 年 1 月第 1 次印刷
印　　刷：上海盛通时代印刷有限公司
开　　本：787mm×1092mm 1/16
印　　张：10.25
字　　数：250 000
书　　号：ISBN 978-7-5669-1962-5
定价（两册）：418.00 元
出版发行：东华大学出版社
地　　址：上海市延安西路 1882 号
邮政编码：200051
出版社网址：http://dhupress.dhu.edu.cn
天猫旗舰店：http://dhdx.tmall.com
营销中心：021-62193056 62373056 62379558
本书若有印装质量问题，请向本社发行部调换。

Encounter
2019 Shanghai Urban Space Art Season Site Project

Edited by: *Encounter* Editorial Board

ISBN 978-7-5669-1962-5

Initiated by: Shanghai Design & Promotion Centre for Urban Public Space
　　　　ARCHIPELAGO
Contributing Editor: XIN Mengyao
Editor: GAO Lulu
Graphic Design: LI Gao

Published in January 2021, by Donghua University Press,
1882, West Yan'an Road, Shanghai, China, 200051.
dhupress.dhu.edu.cn
Contact us: 021-62193056 62373056 62379558